A Geologic Trip across Tennessee by Interstate 40

A Geologic Trip across Tennessee by Interstate 40

Harry L. Moore

OUTDOOR TENNESSEE SERIES
Jim Casada, Series Editor

The University of Tennessee Press / Knoxville

To celebrate Tennessee's bicentennial in 1996, the Outdoor Tennessee Series covers a wide range of topics of interest to the general reader, including titles on the flora and fauna, the varied recreational activities, and the rich history of outdoor Tennessee. With a keen appreciation of the importance of protecting our state's natural resources and beauty, the University of Tennessee Press intends the series to emphasize environmental awareness and conservation.

First Edition.

The paper in this book meets the minimum requirements of the American National Standard for Permanence of Paper for Printed Library Materials. ♾ The binding materials have been chosen for strength and durability.

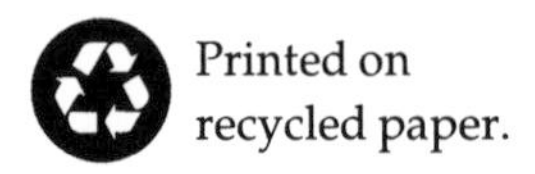

Library of Congress Cataloging-in-Publication Data

Moore, Harry L., 1949–
A geologic trip across Tennessee by Interstate 40 /
Harry L. Moore. — 1st ed.
p. cm. — (Outdoor Tennessee series)
Includes bibliographical references and index.
ISBN 0–87049–832–0 (pbk.: alk. paper).
1. Geology—Tennessee—Guidebooks. 2. Tennessee—Guidebooks.
3. Interstate 40—Guidebooks. I. Title. II. Series.
QE165.M66 1994
557.68—dc20 93-43322
CIP

For my geology mentors:

James Aycock, Garrett Briggs, Don Byerly, Mike Clark, Harry Klepser, Otto Kopp, Larry Larson, Stuart Maher, Robert McLaughlin, David Royster, George Swingle, and Kenneth Walker

Contents

Illustrations

Figures

Maps

Tables

SYMBOL GUIDE TO MAPS

Symbol	Meaning	Symbol	Meaning
75	Interstate Highway		Paved Roads
411	US Highway		Gravel Road
68	State Primary Route		Trail
165	State Secondary Route		Park Boundary
	Mountain		State Line
	City or Point of Interest		Railroad
	State Capital		State Park
	Airport		State Natural Area
P	Parking		Historic Area
	Scenic Overlook	?	Information Center
	Hiking		Horse Stables
	Camping	W	Boat Ramps
	Picnic Area		Cycling

Paleozoic

Pennsylvanian, Mississippian, Devonian-Silurian, Ordovician, Ordovician-Cambrian, Cambrian

Cenozoic

Quaternary, Tertiary

Mesozoic

Cretaceous

Precambrian

Sedimentary/Metamorphic, Igneous/Metamorphic

Preface

An unbecoming avenue of asphalt, Interstate 40 wends its way for mile after dreary mile across the Tennessee landscape. As is the wont of superhighways, in its east-to-west passage I-40 focuses on service rather than scenery, function as opposed to aesthetic fulfillment. In that sense I-40, and for that matter interstates in general, are a sad commentary on the world in which we live. They are sorry symbols of a mad, rushing world, one where haste substitutes for happiness; where beckoning byways are bypassed in an unseemly rush to worship at the altar of material gain.

For the most part, and certainly I-40 is no exception, America's interstates are ugly. They are dual ribbons desecrating the earth. These ribbons are separated by litter-strewn medians planted to non-native grasses, or worse still, by concrete barriers scarring the landscape. To be sure, interstates run straight and true, but they do so only in order to enable haste-driven travelers to speed heedlessly along at 65 miles per hour. Lost as a result of their construction are the slower-paced charms of winding country roads, the endearing appeal of rural America, and the opportunity to pause and reflect in the course of one's travels. Instead, interstates do little more than carry us from one city to another, offering little save cloverleaf gas stations, a myriad of chain motels, and hamburger heavens for havens along the way. What a far, sad cry these modern "conveniences" are from the romance associated with country inns, mom-and-pop crossroads stores, and endearing drive-in restaurants of yesteryear.

Frankly, as I have probably made clear by now, I have a long-standing aversion for interstates. They are robbers of the human spirit, dehumanizing the once pleasant process of travel to such an extent that all that seems to matter once a vehicle is grasped by their tentacles is getting from one place to another. Along with dehumanizing, interstates distort. They have a marked tendency to give us an imperfect picture of the countryside they traverse, offering the wayfarer views of little more than mileage markers, exit signs, and high-rise billboards. For my part, I have long preferred a quiet country road, or even a bustling two- or four-lane highway, to interstates. At least they have character, even if it amounts to nothing more than changing speed zones or the periodic traffic light.

After a first reading of Harry Moore's manuscript, however, my antipathy for interstates, or at least I-40, has undergone a bit of a change. As a one-time resident of Nashville, I have long known I-40 well, with my acquaintance stretching back to those days when the speed limit, at 75 miles per hour, made the landscape even more of a blur. I tolerated I-40 because I really had no choice, knowing that periodically it would offer me the quickest way of finding solace for my troubled soul by taking me back to the high country of the Great Smokies which had been my boyhood home.

Yet each time I traveled I-40 from Nashville to Asheville, it saddened me. The enchanting undulations and majestic hardwood ridges of the Cumberland Plateau received scant visual justice from the fleeting perspective offered by the road, and past Knoxville and Newport, as I moved into the maw of my beloved mountains, things actually worsened. For here I-40 parallels the Pigeon River closely as both cut their way through the main spine of the Appalachians. It is a sordid toss-up as to which is ugliest—the river or the road.

The Pigeon River, stained black by the Champion paper mill in Canton, comes closer to resembling the Styx than should ever be the lot of any earthly waterway. Scudding foam mars its surface everywhere there are rapids, and never is there a rising fish or a happy fisherman to lighten the traveler's heart along the way. As if somehow determined to match the Pigeon's hellish appearance, this particular portion of I-40 also lays bare the earth's soul. Sheer walls of rock, unadorned by any vestiges of vegetation, flank I-40 first on one side and then the other. Frequently garish chain-link fences cling to these barren walls, grim indexes to the instability of the mountains in this part of the Appalachians and stark reminders of the inadequacies of the engineers who planned this portion of I-40. Rock slides remain frequent, and year after year the interstate is blocked by them for hours or even days.

Heading west from Nashville, I-40 again was long an object of my disdain. About the only difference between the journey eastwards and this one was the unalleviated sameness encountered as each passing mile brought me closer to the Mississippi. Where cotton once reigned as king, where mules placidly pulled plows cutting furrows in the dark loam, and where waterfowl once rested in their migrations along the Mississippi flyway, now there was naught but mile after mind-dulling mile of singularly unredeeming landscape.

As I read Moore's words, though, he magically helped transform my views of I-40. I still detest interstates in general, but at least in this one instance he has given me reason to offer a superhighway redemption. In these pages we learn that human beings, through our insatiable lust for progress, sometimes accidentally reminds ourselves of just how short and comparatively insignificant is our role in the sweeping compass of earth's existence. At least that is the thought which came to my mind when pondering ageless geological formations in comparison with the human life span. I for one find such a reminder heartening, because it focuses my attention on the good earth, on the inexorable workings of time, and on the deeper meanings of those transitory moments which are our passage through this world.

As I contemplated Moore's message, another kind of realization slowly dawned. I realized belatedly that "rocks," to put geological formations on the elementary level at which I must confess I've always thought of them, had always been an integral part of my life. As a small boy I marveled at them. Smooth, flat creek stones, sanded and shaped by coursing waters over countless centuries, were tailor-made for service as skipping stones. Other polished pebbles, more round than flat, served as ammunition for my slingshot. It was nice to carry fool's gold in my pocket, dreaming of untold riches as I did so. Other types of rocks and formations—mica, feldspar, and even garnet—likewise received their share of a curious boy's attention.

Then too, I was constantly around rocks, noticing their vagaries of color and infinite variety of shapes, as I waded freestone mountain streams in the delightful quest for trout. As an angler I came to know about deadly acid-bearing rocks which, when exposed as a result of road building or construction, could prove deadly to fish. I knew that there were precious stones, most notably rubies, to be found in parts of the high country, and there had even been productive deposits of gold mined in a few places. Still, considerable potential interest notwithstanding, my knowledge on and thoughts of rock formations were basically characterized by woeful ignorance. I suspect this is pretty typical, and perhaps therein lies the greatest appeal of the present work. The author reminds us, in straightforward language and simple terms readily understood by the uninformed, just how fascinating geology can be when reduced to its basic elements.

He also instructs us, in precisely the painless fashion which is the hallmark of any really informative book, on how exercise in geological exploration can lighten our days and brighten our traveling ways. Thanks to this book, never again will I think of I-40, much less journey along it, in quite the same frame of mind. Moore has given I-40 a new and infinitely more appealing face, one which invites his readers to take a "hands on, eyes on" approach to the land. We normally think of leisurely journeys in terms of stopping to smell the roses, but with Moore that is transformed to pausing to see the rocks.

Motor along in Moore's footsteps, halting when he suggests doing so, and you too will see I-40 from a refreshingly different perspective. We tend to think of the "outdoors" in terms of wildlife and wildflowers, scenic vistas and wild waters. In this work, though, we are given welcome insight into one of the most frequently overlooked aspects of what constitutes complete awareness of the natural world. Namely, to return to my simplistic way of thinking about such things, rocks. The book gives a breadth to the Outdoor Tennessee series through treatment of a neglected subject, and it does so in first-rate fashion. As such, it is a welcome and meaningful addition to the collection.

Jim Casada
Series Editor
Rock Hill, South Carolina

Acknowledgments

When a book takes on a subject as large as the geology of Tennessee, its author soon finds himself indebted to so many people and organizations across the state that only a few of them can be acknowledged here by name. I can hardly begin to thank all the geologists who down through the years have worked so diligently to decipher the geologic history of Tennessee. One geologist in particular who has earned my gratitude not only for technical advice but also for stylistic suggestions is Don Byerly, professor of geology at the University of Tennessee. Also, special thanks to Bob Hatcher, professor of geology at the University of Tennessee, for his comments and suggestions.

Both the National Park Service and the Division of State Parks, Tennessee Department of Conservation, have provided valuable information about their parks, in addition to allowing me to reproduce their maps. The Tennessee Department of Transportation, which has furnished road maps and other valuable highway information, has been most supportive. The Geotechnical Operations Section has kindly provided aerial photographs taken by George Hornal. Ruth Letson of the Tennessee Department of Transportation library helped me find useful information about the history of highway projects in Tennessee.

Special thanks go to Jim Aycock, who gave me encouragement as well as a photograph, and to Nancy Chadwell for typing preliminary drafts of the manuscript. I am grateful for Bettie McDavid Mason's advice and encouragement as well as her editorial skills. I also thank

Jennifer Siler and the staff of the University of Tennessee Press for their patience in seeing me through a second book.

To my mother-in-law, Gertie Richardson, goes my gratitude for her assistance in documenting many side trip routes. As always, I am most thankful for the support of my wife, Alice Ann Moore. Without her guidance, patience, and understanding—especially on weekends and holidays consumed by research and writing—I could never have undertaken this project, let alone completed it.

Introduction

Stretching like a ribbon across Tennessee from the Blue Ridge Mountains in the east to the Mississippi River in the west, Interstate 40 is more than just part of a giant transportation system. Because of the shape of the state and the particular route of I-40 through East, Middle, and West Tennessee—crossing the three grand divisions of the state laterally—this interstate highway provides you with a unique opportunity to view Tennessee's varied landscapes and geology. In fact, by following I-40 across the state, you will not only cross all nine physiographic provinces and sections found in Tennessee but also have the opportunity to see some of the state's most beautiful countryside.

Protrusions of bedrock surfacing as bluffs and natural outcroppings present opportunities for you to look into the geologic past—a history of moving continental and oceanic plates, ancient life forms, and the accumulation and subsequent deformation of ocean sediments. In modern times highways have dissected the landscape, often providing fresh exposures from which to observe and study the bedrock and its structure.

The great valley of East Tennessee, for example, is textured by numerous parallel ridges and valleys that are underlain by folded and faulted rock strata 300 to 500 million years old. In contrast, the gently rolling to flat landscape of West Tennessee is underlain by silt, sand, and clay, some less than a million years old.

On your geologic trip across Tennessee, you will see all three of the basic types of rocks: igneous, sedimentary, and metamorphic. While using I-40 as your main reference line, you may be tempted to stop

Fig. 1. The mountainous Blue Ridge Province in East Tennessee contains the oldest rocks found in the state; highest peak in middle background is Thunderhead Mountain (5,530 feet), Great Smoky Mountains National Park.

along the interstate to look at the rock exposures, but you must resist the urge. *Stopping along the interstate is extremely dangerous and is against the law except for emergencies or at designated rest areas.*

In order for you to have opportunities to get out of your car—to see and touch the rocks—nine side trips are provided along with the annotated road log for I-40 itself. These side trips will take you along scenic routes through rural sections of Tennessee to explore the landscape, examine rock formations, and, if you choose, do some hiking and enjoy the outdoors on your geologic trip across the state.

A State and a Highway

Part 1 of the book gives you some background information about what you will be seeing as you follow the road log for I-40 and the side trip itineraries in Part 2.

"Oceans, Rocks, and Time" presents some of the general concepts and basic principles of the science of geology. Here you will find

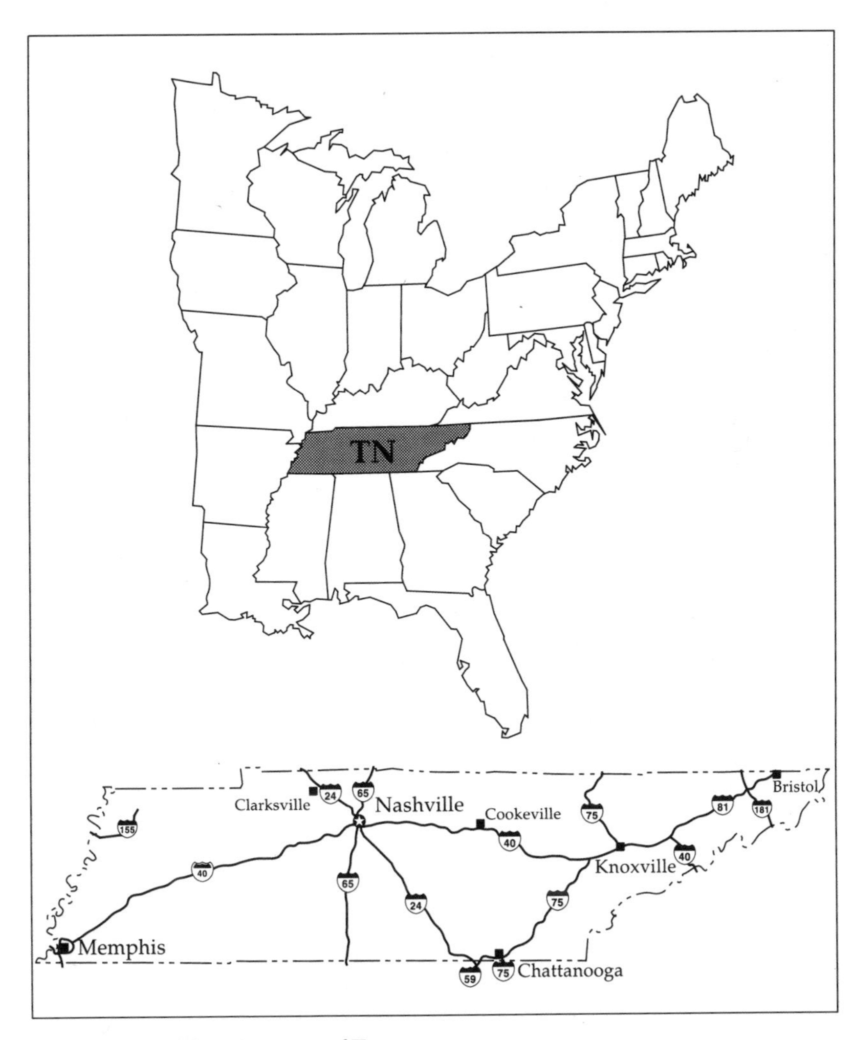

Map 1. General location map of Tennessee.

Fig. 2. Waterfalls abound in the East Tennessee landscape, reflecting the occurrence of massive beds of hard bedrock. Waterfall is Rainbow Falls, Great Smoky Mountains National Park.

discussion of topics such as elements, minerals, rocks, the rock cycle, continental drift and mountain building, geologic time, and earth's past life to prepare you for the next chapter.

"Tennessee's Geologic History," focusing upon the general geologic character of the state, is organized around the major eras of geologic time and how those eras are represented in the rocks found across Tennessee. In addition, this chapter shows you the relationship between the physiographic landforms of the state and their underlying geology.

"Tennessee's Topography" introduces you to the major physiographic provinces forming the landscapes across the state. Each of the nine physiographic provinces—the Blue Ridge, the Valley and Ridge, the Cumberland Plateau, the Eastern Highland Rim, the Central Basin, the Western Highland Rim, the Western Valley of the Tennessee River, the Gulf Coastal Plain, and the Mississippi River Alluvial Floodplain—is described and located. You are provided with details about each of these physiographic areas as well as with descriptions of topographic areas of lesser magnitude.

"Geologic Environmental Issues" identifies and describes some of Tennessee's environmental problems that have geologic origins or are controlled or influenced by geologic conditions. Among the topics discussed are landslides, sinkholes, groundwater pollution, solid waste disposal, and mining, as well as engineering geology along highways.

"The Making of a Highway" offers a nontechnical explanation of how a highway is built. In this chapter, you will learn not only about the mechanics of locating, designing, constructing, and maintaining a highway, but also about the environmental considerations and geological problems involved, with some specific examples from the history of I-40 in Tennessee.

Road Log, with Side Trips

Part 2 of the book provides a guide to Tennessee, using I-40 as your basic route (maps 2 and 3). In addition to the road log for I-40 itself, there are side trips to beautiful and geologically interesting areas. These side trips, at various points along I-40, are usually within an hour's drive (or less) from the interstate.

The comprehensive road guide to I-40 begins at the North Carolina–Tennessee state line and goes west across the state, ending at the

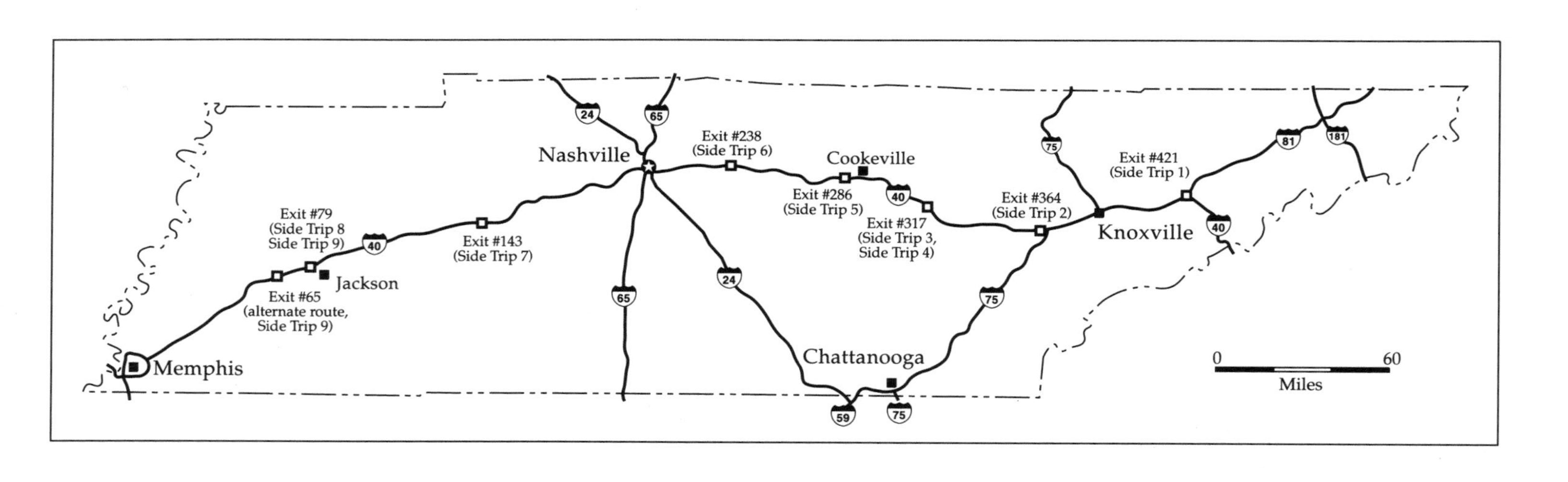

Map 2. Interstate-40 across Tennessee, showing selected exits. After TDOT Official Highway Map.

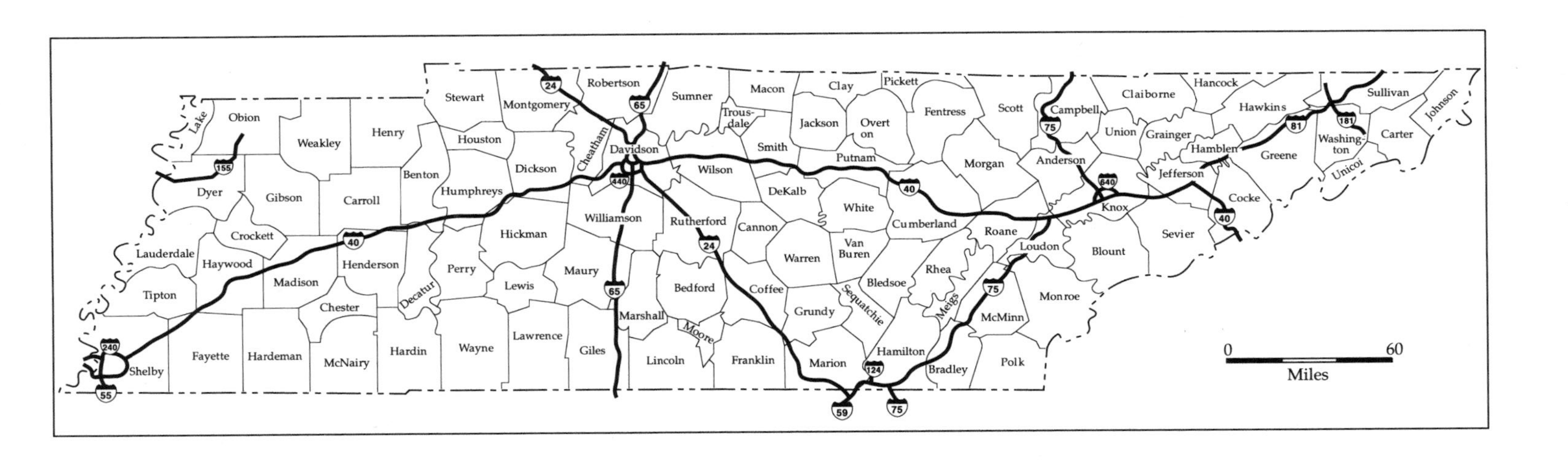

Map 3. County outline map of Tennessee showing interstate highways. Courtesy of TDOT Bureau of Planning and Development.

Tennessee-Arkansas state line (the Mississippi River) at Memphis. Geologists usually describe the geology of an area by beginning with the older strata and progressing to the younger strata, as the younger rocks are composed of fragments of older ones. The east-to-west direction was chosen because the rock strata become younger from east to west, the oldest rocks being found in the mountains of the Blue Ridge and the youngest strata in West Tennessee.

Highway Signs

Interstate highways as well as all federal highways and all state primary and secondary roads are marked with identifying signs. The interstate system utilizes a red, white, and blue shield with the interstate numerals across the face of the shield (Fig. 3). Every federal (U.S.) highway is marked with a black-and-white shield, which has the number of the particular highway on it (Fig. 4). These roads usually run across state lines—for example, U.S. 441 or U.S. 70.

The Tennessee state highway system includes both primary and secondary roads. A primary highway, marked by a rectangular, black-and-white sign with the highway number inside the rectangle, is

Fig. 3. A typical interstate highway marker.

Fig. 4. U.S. highways are indicated by a white shield with black numbers.

usually a main route connecting cities and towns (Fig. 5); in some cases, it is simultaneously a U.S. highway. A secondary highway, identified by a triangular, black-and-white sign with the highway number inside the triangle (Fig. 6), is a road carrying a smaller volume of traffic and is usually located in rural areas; secondary highways connect to primary highways. County and urban roads are not identified.

Fig. 5. Tennessee primary highways are marked with a white rectangle sign containing black numerals and the word "Tennessee." Some of the more scenic Tennessee routes are indicated with the Tennessee scenic parkway sign containing an image of the state bird—the mockingbird.

Fig. 6. Tennessee secondary highways are indicated by a white triangular sign containing black numerals.

Odd-numbered highways run north-south; even-numbered ones run east-west. Some U.S. routes are split, with one route going east and the other west (or one north and one south). A good example is U.S. 11. North of Knoxville, U.S. 11 is split into U.S. 11E and U.S. 11W, both of which run north-south. The two routes rejoin to form U.S. 11 again at Bristol.

All state primary and secondary roads and interstates are numerically marked every mile. These marks, called milepost numbers, begin with 0.0 and progress upward until a county boundary is crossed. The milepost numbers run from south to north or west to east, depending on the route. For example, S.R. 14, an east-west highway, begins with Milepost 0.0 at the Mississippi-Tennessee state line; the numbers on the milepost markers increase as the route goes east to the Shelby-Tipton county line. When S.R. 14 reaches that boundary, the milepost number restarts with 0.0 and increases again across Tipton County. The same is true for north-south roads, such as S.R. 1 (also identified as U.S. 11W) in Knox County or S.R. 111 in Van Buren, White, and Putnam counties.

The milepost signs—small, rectangular, green signs with white numbers—are usually found along the side of the highway on the shoulder at one-mile intervals (Fig. 7). In the state highway system, the milepost signs have a smaller white number at the bottom of the sign (just beneath the mileage number) indicating the state route number of the road. Along the interstate highways the milepost signs are also green, but they are much larger and contain just the mileage number (Fig. 8). The signs, placed along the shoulder of the road at one-mile intervals, are present on both sides of these divided highways.

Each interstate highway has its own milepost numbers. In Tennessee, I-40 begins with 0.0 at the Tennessee-Arkansas state line and progresses eastward to Milepost 451, the last one before the Tennessee–North Carolina state line (the actual mileage to the state line is 451.8). Remember, I-40 is an even-numbered highway—that is, east-west—and in Tennessee the mileposts run from west to east on even-numbered highways.

Identifying the milepost markers along the highways is very important in using this guide. The Road Log descriptions identify such elements as rock exposures, landscape changes, political boundaries, and rivers by referencing the milepost numbers along the highway.

Because of the geological considerations mentioned earlier, the Road Log descriptions begin in the east (at Mile 451.8) and proceed westward. Accordingly, the mileages are identified as though you were

Fig. 7. Every Tennessee state route is marked with small milepost signs at one-mile intervals. The larger number is the mile designation; the smaller number indicates the state route. This sign indicates Mile 4 on S.R. 73.

Fig. 8. Milepost markers are located along the interstate shoulder at one-mile intervals. These signs are green with white numerals.

traveling west on I-40, and the numerals decrease rather than increase. If you are an eastbound traveler, you will need to make log mile adjustments as necessary. Any exit number, however, is derived from the milepost at which the particular exit is located and is thus the same whether you are going east or west on I-40.

The mileages stated in the road log are to the nearest 0.1 mile. You can expect to have some very slight variations in mileage because not all vehicles have their odometers calibrated the same way. All milepost signs are installed and maintained by the Tennessee Department of Transportation. Because of highway construction projects, or traffic accidents, however, some individual milepost signs may be missing from time to time.

Road Log

The road log includes material on not only the geologic conditions but also other scenic and geographic points of interest along I-40 across Tennessee (Fig. 9). Physiographic provinces, types and ages of rock formations, fossil and mineral occurrences, county and city boundaries, and notable landmarks such as state parks, historic points of interest, and streams and mountains are noted in the road log for easy identification (Fig. 10).

The road log begins at the North Carolina–Tennessee state line and proceeds westward to the Mississippi River at Memphis. The three grand

Fig. 9. Interstate 40 winds across the Tennessee landscape, providing the traveler an opportunity to see the state's varying landscape. This view is to the east as I-40 traverses the Cumberland Plateau. TDOT photo by George Hornal.

Fig. 10. As you travel the Tennessee landscape exploring the scenic byways, you will experience some of the state's most beautiful and interesting areas.

divisions of the State of Tennessee—West Tennessee, Middle Tennessee, and East Tennessee—are clearly indicated in the road log, along with the physiographic provinces and subprovinces found within each.

Another feature is the inclusion of environmental topics related to engineering geology problems along I-40. At the appropriate mileages the road log mentions such geological problems as landslides and sinkholes and provides details about their occurrence and eventual correction—reminders about the complexities of the environment and how humans in their quest to manipulate the land surface for their benefit have had to deal with unique geological conditions.

"East Tennessee" (Miles 451.8–297.0) presents the most dramatic part of the Tennessee landscape. Natural rock bluffs, rock shelters, waterfalls, parallel ridges and valleys, and high mountains are among the features described in this section of the road log. The three physiographic provinces described in this section are the Blue Ridge (Miles 451.8–443.0), the Valley and Ridge (Miles 443.0–341.5), and the Cumberland Plateau (Miles 341.5–297.0). You are introduced to varied geologic conditions consisting of flat-lying to folded and faulted strata of the Paleozoic and Precambrian ages. The four side trips in this section reflect this geologic diversity. Side Trip 1 to Cumberland Gap National Historic Park, which takes a geologically interesting route via Clinch Mountain, shows how geology plays a crucial role in history: early settlers used this natural gap in the mountains (the boundary of two physiographic provinces, the Valley and Ridge and the Cumberland Plateau), and it was also important strategically during the Civil War; today the state lines of Tennessee, Kentucky, and Virginia come together at this point. Side Trip 2 to Cades Cove, one of the loveliest areas in the Great Smoky Mountains National Park, via Townsend, Tennessee, features two of the unusual valleys in the Blue Ridge province: Tuckaleechee Cove and Cades Cove; in these cove areas erosion has weathered through the older Precambrian rocks and formed "windows" into the younger Paleozoic rocks exposed. Side Trip 3 to Twin Arches, in the Big South Fork National River and Recreation Area (north of Crossville), provides another spectacular example of Tennessee's geologic formations; these two huge arches, end on end, form a gigantic natural bridge complex. Side Trip 4 to Fall Creek Falls State Park (south of Crossville), which also provides a look at the beautiful Sequatchie Valley, focuses on Pennsylvanian Age sandstone and shale, over which pours the highest waterfall in the eastern United States.

"Middle Tennessee" (Miles 297.0–135.0) treats the rolling and hilly landscape in the center of the state. The three physiographic provinces included in this part of the road log are the Eastern Highland Rim (Miles 297.0–272.5), the Central Basin (Miles 272.5–186.0), and the Western Highland Rim (Miles 186.0–135.0). In this section of the state, you encounter flat-lying to gently dipping limestone and shale of the Paleozoic Age. Side Trip 5 visits Burgess Falls State Natural Area, near Cookeville, to see some lovely cascades over Mississippian age limestone. Side Trip 6 is to Cedars of Lebanon State Park, a beautiful cedar glade near Lebanon. Side Trip 7 includes both the Wells Creek structure, a fascinating cryptoexplosive structure near Cumberland City, and Dunbar Cave State Natural Area, near Clarksville.

"West Tennessee" (Miles 135.0–0.0) details the flat-to-rolling landscape of the western third of the state (Fig. 11). The three physiographic provinces in this section are the Western Valley of the Tennessee River (Miles 135.0–133.0), the Gulf Coastal Plain (Miles 133.0–1.1; Fig. 12), and the Mississippi River Alluvial Floodplain (Miles 1.1–0.0). Here you encounter Mesozoic and Cenozoic Age rock formations consisting of sand, silt, clay, and marl. Side Trip 8 goes to Reelfoot Lake State Park to see the famous lake created by violent earthquakes in 1811 and 1812. Side Trip 9 visits the Fort Pillow State Historic Area to view the Chickasaw Bluffs and the Mississippi River floodplain.

Fig. 11. West Tennessee is characterized by its relatively flat topography. Note how flat the horizon is in this West Tennessee sunset.

Fig. 12. Tennessee is bounded on the West by the Mississippi River.

Side Trips

You can have a geologically interesting trip simply by following I-40 across Tennessee, and it is arguably the most scenic interstate in the whole system. But if you would like to get a really close look at some rock formations—and have more leisure to enjoy the scenery and local culture of the state—you should try to take one or more of the nine side trips, with optional hiking trails, noted above as well as at the appropriate spot in the road log and listed in tabular form for your convenience (Table 1). Maps guiding the reader to each natural area are accompanied by a descriptive narrative.

Each side trip begins with information about the destination, the route (including mention of what cities and towns are along the way), the length of the trip, the nature of the roads, the beginning and ending elevations, special features, an optional hiking trail, available U.S. Geologic Survey maps, and a brief list of relevant terms for reference through the index and glossary. Specific directions to each destination from the specified I-40 exit are given. Following the directions is a general description of the route, which includes geographic, geologic, and historical details of interest.

Table 1. Side Trip Locator

Side Trip	I-40 Exit	Destination
1	421	Cumberland Gap National Historic Park
2	364	Cades Cove, Great Smoky Mountains National Park
3	317	Twin Arches, Big South Fork National River and Recreation Area
4	317	Fall Creek Falls State Park
5	286	Burgess Falls State Natural Area
6	238	Cedars of Lebanon State Park
7	143	Wells Creek Structure and Dunbar Cave State Natural Area
8	79	Reelfoot Lake State Park
9	79	Fort Pillow State Historic Area

All mileages cited in the side trip descriptions are Log Miles, with the beginning point (0.0) at the place where you turn off the I-40 exit ramp onto a state or federal highway and begin the side trip. The Log Miles run consecutively from 0.0 to the destination of the side trip (in many cases, a hiking trailhead). One exception is the Cades Cove side trip, whose Log Mile total ends at the beginning of the one-way loop road, which has its own road log mileage.

Please note that the Log Miles are not to be found on any external signs along the side trip routes. You will need to keep up with the odometer mileage yourself if you wish to have an accurate location of points detailed in the side trip descriptions.

The hiking or walking trails are provided so that you can see Tennessee geology at close range as well as have an opportunity to get out into the countryside. These trails vary in distance from one-fourth mile to over eight miles round-trip. Scenic mountain coves, caves, rock pinnacles and arches, waterfalls, and lakes are just some of the natural features of the landscape for you to enjoy. At the end of each hiking trail description is a list of addresses and telephone numbers that you may wish to consult for additional information, such as lodging, food, and schedules of events in the area. If your interests include spelunking or rock collecting, you can turn to the appendices for the names and addresses of caving clubs and rock and mineral clubs in East, Middle, and West Tennessee.

As you prepare for your side trip, you may also wish to obtain additional (and perhaps more technical) information about your destination. Where appropriate, some of this information is included in parenthetical references within the narrative; you can find the specific details about each of these entries in the reference list at the back of the book. The Tennessee Division of Geology, state colleges and universities, and your local library are good sources for finding this additional information. A Tennessee highway map will make a good companion for this guidebook. Free copies of maps are usually available at welcome centers along Tennessee interstates.

Be sure to remember that except for the designated rest areas, stopping along the interstate highway can be hazardous and is against the law unless you have an emergency. For this reason, please use the side trip routes for looking at (or collecting) rock and fossil specimens. *Remember that collecting specimens is not permitted in state and national parks.* Study and enjoy the rocks and living organisms, but leave them where you find them so that others will be able to have the same privilege.

PART 1

A State and a Highway

Oceans, Rocks, and Time

Mountains, hills, and valleys; rushing streams and waterfalls; vast rivers and lakes; caves and coves; swamps and mysterious craters—all, in some way, are a part of our landscape in Tennessee. Over many millions of years, the forces of land upheaval and erosion have sculptured the land surface into a mosaic of topographic forms. Even earlier, cosmic forces were at work developing the framework for what would become our solar system and our planet earth.

In the far distant past, our solar system probably had its origins in a mass of cosmic dust and gas left over from the Big Bang that created the known universe. Turbulence within this cosmic cloud—perhaps from a nearby exploding star (supernova)—caused the cloud to contract. As the cloud of dust and gas (solar nebula) contracted, its rotation caused it to assume the shape of a disk. Gravity caused the nebula's particles to attract each other so that it contracted even more. The contraction increased the density of the gas and dust, and that, in turn, raised its temperature to tens of millions of degrees so that hydrogen fusion began (Glass 1982: 5). Our sun was born as a protostar.

During this process some of the gas and dust remained in the rotating disk around the protostar. Through localized processes of gravitational attraction and collision, the dust and gas in the outer disk accumulated in bodies that eventually reached sizes to become our planets and moons.

Because the heat and pressure generated by our sun affected nearby objects more than distant ones, the planets developed into two types: the inner terrestrial planets and the outer, gigantic, gaseous planets. The terrestrial planets are the result of the formation of metal oxides and silicates in the area of high temperature and pressure nearer the central sun. Further out, the temperatures became lower and lower, allowing the ices of water, carbon dioxide, methane, and ammonia to stabilize along with the iron and silicates. Where the temperature was low enough for large amounts of the lighter gases to accumulate, the outer planets grew to much larger sizes than the terrestrial planets (Glass 1982: 6).

The inner terrestrial planets are known as Mercury, Venus, Earth, and Mars. The outer giant planets are Jupiter, Saturn, Uranus, Neptune, and Pluto. In between the inner and outer planets is a band of broken rock debris called asteroids.

Our planet, Earth, is a very unusual place in this solar system. Not only is it the largest terrestrial planet, but its satellite is larger in proportion to the size of the planet than any other in the solar system. In addition, Earth is the only planet with large amounts of liquid water on its surface, and it is the only planet where (so far as we know) life has developed (Glass 1982: 42). Even more interestingly, it is the only planet where life has developed to a level that it can ponder about its origins.

The Earth, Its Origins, and the Continents

The protoearth is believed to have been an accumulation of cold rock debris that later heated up through the continued effects of additional impacting rock fragments, gravitational compression of the entire mass, and radioactive decay (Tozer 1977: 2, 23–26). Due to the heating of the earth, the rock material became fluid enough that its components differentiated into layers of similar material. As a result, the earth is composed of a metallic iron core, an iron and magnesium silicate mantle, and an outer, silicate-rich crust called the lithosphere.

The movement of the material deep within the earth influences other factors on the planet. The outer part of the core is thought to be more liquid than the inner core. Convection of the outer core is thought to be responsible for earth's magnetic field (Kopal 1973: 32). Convection of the more viscous upper mantle, the asthenosphere, is believed to be responsible for the driving mechanism of plate tectonics and continental drift.

Initial differentiation of substances and simultaneous cooling of the protoplanet resulted in the formation of several large crustal bodies called cratons. Theorists believe there were about a dozen or so crustal plates that fit together in a mosaic-like form on the earth's surface (J. T. Wilson 1963: 41–45). The plates were separated by a series of deep oceanic trenches and shallow oceanic ridges.

The rock material composing the continental (cratonic) plates is less dense than the ocean plates, which are composed primarily of basalt and similar rocks. The ocean and continental plates form the lithosphere of the crust and rest ("float") on the asthenosphere.

Convection currents (the rise of material as it is heated and its subsequent sinking as it cools) in the upper mantle cause upwellings and downward-moving zones within the asthenosphere material. Where crustal plates are attached to these moving zones in the asthenosphere, the crust tends to move. Where upward-moving currents occur in the asthenosphere, there is a corresponding midoceanic ridge in the crust (for example, the Mid-Atlantic Ridge in the Atlantic Ocean). Where there are downward-moving zones in the asthenosphere there are downward-moving zones in the crust, called trenches (Fig. 13). (One example is the Mariana Trench in the Pacific Ocean.) The deep ocean trenches can occur along continental margins and volcanic island arches. The continents tend to be pulled and pushed by these convection currents.

During our geologic past, the continents have come together and moved apart several times, leaving mountain chains in their wake. The Himalayas, Andes, Alps, Rockies, and Appalachians are all examples of mountain chains that have resulted from these continental collisions.

Sediments that formed in the oceans between the continents later became part of the continents when they were sutured together as the continental plates collided. The deformed rocks that compose the mountain chains are partly those same originally flat-lying sediments that formed in the intervening ocean basins and continental margins.

Rifting (splitting apart) of the North American continent during the late Precambrian eon initiated a sequence of sedimentation and crustal movements that resulted in the formation of much of the rocks currently found in the Blue Ridge and Valley and Ridge of East Tennessee (Hatcher and Lemiszki 1991).

The gradual collision of the African Plate with the North American Continent—a process that took perhaps 200 million years—ended approximately 250 million years ago, and resulted in the formation of the Appalachian Mountain chain. The compressive forces generated by

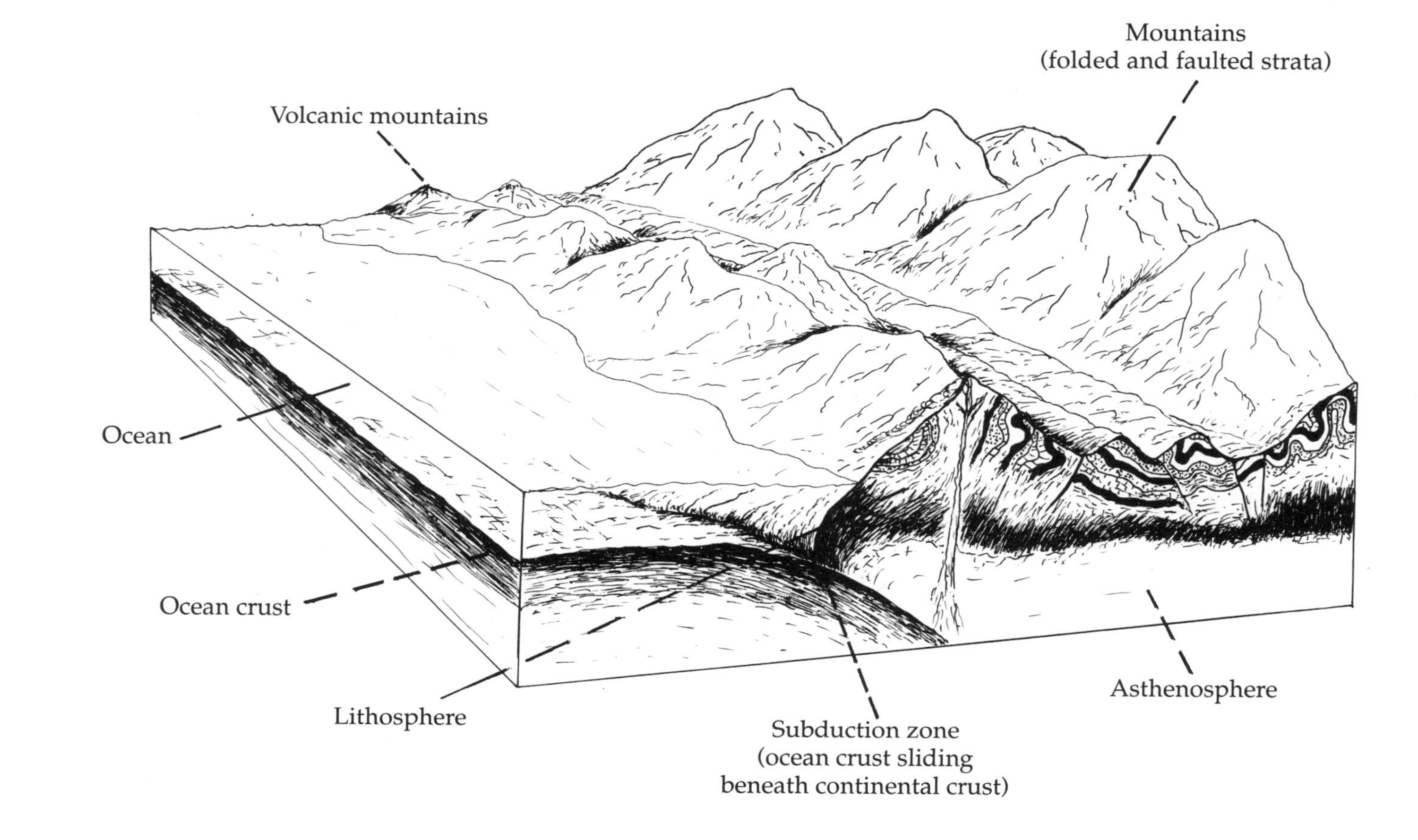

Fig. 13. A schematic drawing of an oceanic plate subducting beneath a continental plate.

the two colliding continental plates deformed the rock layers into belts of parallel folds, which you can see today in the Valley and Ridge province of Tennessee, Virginia, and Pennsylvania, and a large zone of complex deformation known as the Blue Ridge of Tennessee and the Blue Ridge and Piedmont of North Carolina and Virginia.

Minerals

The fundamental building blocks of the rocks we see are elemental atoms (such as hydrogen, oxygen, and carbon). Large groups of like elemental atoms form elemental minerals such as gold, silver, and copper. More complex arrangements of atoms form mineral compounds such as an element plus oxygen (SiO_2, silicon dioxide or quartz; Fig. 14) or an element plus oxygen and carbon ($CaCO_3$, calcium carbonate or calcite).

Geologists describe such combinations as minerals: naturally occurring substances that have definite atomic arrangement and distinctive physical properties and are a product of inorganic processes. Some of the more common minerals found in Tennessee include quartz, calcite, hematite, pyrite, feldspar, mica, limonite, dolomite, fluorite, barite, sphalerite, and lead. A geologist checks certain physical

Fig. 14. Quartz (SiO_2), one of the more common minerals on earth, is a major constituent of Tennessee rocks.

properties to ascertain the correct mineral identification. Of all the physical properties that are checked, hardness is one of the most widely recognized and often used. Mohs Scale of Hardness, the most used series, runs from 1 (softest) to 10 (hardest); each mineral in the scale will scratch all those whose hardness number is lower (Table 2).

A mineral's cleavage (the tendency to break or split along certain defined smooth planes) and structure (whether crystalline or amorphous) are used just as commonly as hardness by geologists to identify minerals in the field. However, laboratory methods can be more exact and include chemical tests and analysis, microscopic identification of a mineral's optical character, and even X-ray diffraction "fingerprinting" of minerals.

Kinds of Rocks

What we see when we examine the earth's crust are not just minerals, but aggregates of minerals. We can define a rock as simply an aggregate of minerals. Most rocks are composed of two or more minerals, such as granite (which may be composed of quartz, mica, and feldspar in combination). However, some rocks are composed primarily of just one mineral, such as limestone (calcite) and quartzite (quartz).

In studying the earth's crust, geologists have divided rocks into three basic categories according to their origins: igneous, sedimentary, and metamorphic. All three rock types are found in Tennessee, but sedimentary rocks are by far the most common.

Table 2. Mohs Scale of Hardness

Mohs Number	Mineral
1	Talc
2	Gypsum
3	Calcite
4	Fluorite
5	Apatite
6	Feldspar
7	Quartz
8	Topaz
9	Corundum
10	Diamond

Igneous Rocks

The origin of igneous rocks is molten material (magma) in the interior of the earth. Molten rocks that burst to the surface and cool rapidly are called extrusive igneous rocks. Igneous rocks whose molten material rises but cools slowly beneath the surface of the earth are called *plutonic* or intrusive. Because the rate of cooling governs the size of crystals, extrusive igneous rocks are generally fine-grained and intrusive ones coarse-grained.

Extrusive igneous rocks, commonly called *volcanic* rocks, consist of lava flows, volcanic ash, and volcanic breccia. Intrusive igneous rocks appear mainly as large granite bodies several miles across or as thin, tabular bodies called dikes and sills. A dike results from the intrusion of magma into a vertical fracture in rock layers; a sill, in contrast, is produced when a sheet of magma intrudes between horizontal or nearly horizontal layers of rock.

In Tennessee igneous rocks are found at the surface in the middle and upper portions of extreme East Tennessee (Blue Ridge province). Exposures of Beech Granite can be seen in Johnson, Carter, and Unicoi counties. Some smaller thin bodies (two to six feet thick) of tabular intrusive dikes are also found in Carter and Unicoi counties and in some regions of the Great Smoky Mountains.

Volcanic ash deposits are thought to be the origin of several bentonite clay layers found in Ordovician rocks of Tennessee. Located just northeast of Laurel Bloomery in Johnson County are volcanic rocks of Precambrian age (Mount Rogers Group). Exposed in this area are massive lavas and tuffs (rocks formed from compacted volcanic fragments). There are also exposures of very fine-grained granites as well as the tuffs and lavas; all have been metamorphosed to varying degrees.

Sedimentary Rocks

Sedimentary rocks are those which form from the material deposited by water, wind, ice, and organic sources. Sedimentary rocks, covering most of Tennessee's landscape, form from accumulations of sediment in the oceans, rivers, lakes, ponds, swamps, and deserts. One of the distinguishing characteristics of sedimentary rocks is their stratification, the sorting of sediment into layers.

When geologists refer to stratified rocks, they use the terms *bed* and *bedding plane* to describe the layers of rock. In general, each bedding plane marks the termination of one deposit and the beginning of another. A bed is the layer of rock between two bedding planes. In the absence of precise and universally accepted definitions of these general categories of bedding, here are some guidelines about thickness: thin beds are an only inch or two (or less), medium beds vary from about two inches to about one foot, and thick beds vary from over a foot to ten feet or more.

Geologists study the beds of stratified rock to understand the geologic structure and even the geologic history of those rocks. By measuring the directional features of a bed of rock, they are able to discern the structure of rock strata at and just beneath the earth's surface. Through comparisons of the various beds, they can reconstruct the history of how the rock layers were formed.

Most sedimentary rocks are made up of fragments of preexisting rocks. Quartz sand derived from the weathering of granite, for example, results in the formation of sandstone. However, some sedimentary rocks (most notably limestone, dolostone, calcareous shale, carbonaceous shale, and coal deposits) are formed from animal and/or plant deposits.

In Tennessee sedimentary rocks comprise most of the landscape we see. Limestone characterizes the flat to rolling landscape of the Central Basin and Highland Rim areas, sandstone and shale make up the caprock for the Cumberland Plateau, and sequences of folded limestone, shale, and sandstone comprise the rocks of the Valley and Ridge in East Tennessee.

West Tennessee is also composed of sedimentary deposits. However, most of these deposits (sand, silt, clay, and marl) are not lithified into solid rock layers. They are, just the same, sedimentary in origin and classified as sedimentary rocks.

Metamorphic Rocks

Metamorphic rocks are those that have formed from preexisting rocks, either sedimentary or igneous, due to alteration by heat, pressure, and/or chemical change (Table 3). Some of these rocks may have been only slightly altered and maintain their original character (a metasiltstone, a metasandstone, or an argillite, for example). Others may be so altered that their original origin may be difficult, if not impossible, to ascertain (schist and gneiss).

In Tennessee most of the metamorphic rocks are in the Blue Ridge province of extreme East Tennessee. Slightly altered shale (argillite and slate), and sandstone (metasandstone) comprise a large part of the metamorphic rocks in the Great Smoky Mountains National Park. Highly altered rocks such as phyllite, gneiss and schist are found in Unicoi, Carter, and Johnson counties.

Table 3. Some Common Metamorphic Changes

Original source rock	Typical resulting metamorphic rock
Sedimentary	
Sandstone	Quartzite
Shale	Slate, phyllite, schist, argillite
Limestone	Marble
Igneous	
Light-colored (granite)	Gneiss
Dark-colored (basalt, diorite)	Schist, slate, serpentine

Formations

In studying the rocks of Tennessee, geologists have been able to differentiate the rocks of the state into groups or formations (for example, the Knox Group or the Hermitage Formation). A formation name is simply a convenient means by which geologists can refer to a mass of rocks that have like characteristics and can be mapped at the earth's surface. Rock formations are generally rock units of sufficient size and thickness to be mapped on a scale such as the 1" = 2,000' standard employed in United States Geologic Survey 7 1/2 Minute Quadrangle Maps.

The name applied to a rock formation generally comes from the geographic area where it was first mapped. The name usually comes from a geographic feature such as a mountain or river, or a community name. The kind of rocks found in a given formation are usually the same; however, rock units can range in type from one place to another. For example, the Ottosee Formation found in East Tennessee may consist of a gray shale in one locale and a limestone a short distance away. In addition, the color of a rock formation may vary as well; indeed, since color is site specific for rocks, it should not be used as a guide for identifying formations. Some rock formations are varicolored, ranging from brown and gray to maroon and green (for example, the Pennington Formation, Road Log, Mile 296.0).

Structures

In our discussion of the geology of Tennessee, some additional terms must be defined. Structural features of the earth's crust provide dramatic texture to the lithosphere (Fig. 15).

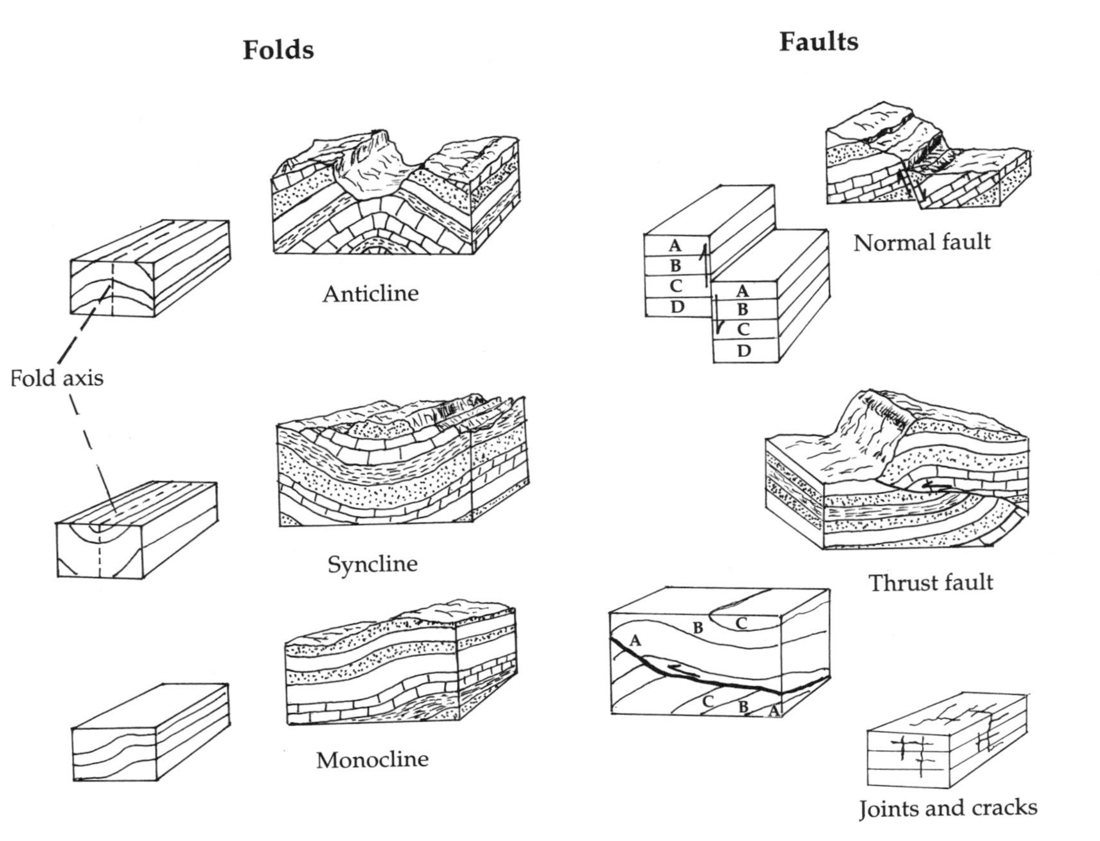

Fig. 15. Typical geologic structures in Tennessee rocks.

Folds

Rock strata are commonly folded into warp-like structures as a result of the moving and colliding crustal plates (Fig. 16). An anticline is an upward bend or warp in which the older rocks are in the center (easily observed in layered rocks). The opposite, a downward bend or downwarp, is called a syncline, in which the youngest rocks are in the center. Variations in the folds produce such structures as asymmetrical, monoclinal, and chevron folds.

Faults

When the pressure on the rock is great enough, the rock will usually break. A fault is a break in a rock mass along which movement has taken place. A thrust fault is produced when one rock mass is pushed over another body of rock at a low angle; generally the overriding rock mass is older than the overridden one (Fig. 17). Most of the faults in East Tennessee are thrust faults. A normal fault is a fault with steep fault planes. In Middle Tennessee normal faults can be seen in the numerous interstate road cuts around Nashville.

Fig. 16. Folded rock strata are common in East Tennessee and reflect the deformation of the strata that occurred over 200 million years ago.

Fig. 17. This exposure of a thrust fault in Grainger County clearly shows the fault plane (dark diagonal line that dips from left to right).

Fractures

In your travels across Tennessee you can see it is obvious that the rock strata break or fracture along definite planes. In fact, these planes of broken rock dominate most rock exposures across the state. These fractures provide the main avenues by which the rock strata are weathered, broken apart, and eroded away. Two major kinds of fracture are common in the rocks of Tennessee: jointing and cleavage.

Jointing. The most simple kind of a fracture is a joint, a fracture in which the breaks are more or less perpendicular to bedding and have shown no appreciable movement along the fracture. Joints usually occur in groups or sets in which a repetitive series of breaks may be found (e.g., cracks spaced evenly at three-foot intervals). Most of the layered rocks, easily observed along I-40, exhibit numerous joints. Particularly good examples of joints can be seen along the trail to the base of Fall Creek Falls, where some of the joints extend vertically over 70 feet and horizontally over several hundred feet (see Side Trip 4 to Fall Creek Falls State Park). In crystalline rocks there is a form of jointing called *sheeting*—fractures with no slippage that result typically from pulling apart (tension) or contraction.

Cleavage. A second kind of fracture that is somewhat more complex is known as cleavage. Cleavage fracture is the tendency of a rock mass to break or split along closely spaced planes that are usually inclined to the bedding planes of strata. When viewed in an outcrop, the cleavage fractures appear as cracks that tend to develop a diagonal pattern between layers of rock. Cleavage fracture can be seen in the rock exposures along I-40 between Miles 451.8 and 447.0 (East Tennessee, Blue Ridge province).

Cleavage, which generally develops in response to pressure, is often associated with metamorphism, folding, and faulting of rocks. Several types of cleavage, including fracture cleavage and slaty cleavage, can occur in rock masses. Cleavage fracture usually is well developed in rock types such as shale, phyllite, slate, and metasiltstone.

Time

The element of time is foremost in the minds of humans. We need time to clean the house, time to finish the report for class, more time to complete the project at work. We are constantly being asked to hurry up—hurry and finish the report, hurry and complete the project. Hurry, hurry, hurry! Time, time, time!

Geologists also speak of time in their unraveling of earth's history. Geologic time, however, is not mere seconds, minutes, hours, days, weeks, and years. Geologic time consists of vast amounts of time—hundreds of millions, and billions of years. The average life of a human (say 75 to 80 years) is but just a blip of a second in geologic time.

Our earth is believed to be over 4.5 billion years old. We humans think in terms of our 75-to-80-year life spans, but how can we think in terms of millions of years? Perhaps we can reduce the number of a million to a level most of us can comprehend. Let's start with seconds. A million seconds is equal to 11.57 days. A million minutes is equal to 694.44 days. A million hours is equal to 41,666.67 days, or about 114.15 years.

Most of us will not live a million hours. The United States has been a nation only 7.8 percent of a million days. What we call civilization has been around for less than 10,000 years or 1 percent of a million years. Any block of time that we can easily associate with recorded human history becomes insignificant when compared to the millions or even hundreds of millions of years in which geologists think.

Geologists divide time into four basic segments. Listed from oldest to youngest, they are the Precambrian eon, the Paleozoic era, the

Mesozoic era, and the Cenozoic era (Table 4). All four are represented in the geologic history of Tennessee and will be described in more detail in the next chapter.

Through processes of deductive reasoning, scientific study, and the understanding of the decay of radioactive elements, geologists now believe that our planet earth was formed about 4.5 billion years ago. From that beginning to about 570 million years ago is a span of time (some 4 billion years) that geologists call the Precambrian Eon. It is during the middle to later portions of Precambrian time that primitive life is thought to have begun. Rocks of Precambrian age are found in Tennessee only in the Blue Ridge province of extreme East Tennessee.

The next oldest segment of time is the Paleozoic ("ancient life"). The Paleozoic era began about 570 million years ago and lasted approximately 340 million years, ending about 230 million years ago. The seven periods of time that make up the Paleozoic era, from oldest to youngest, are the Cambrian, Ordovician, Silurian, Devonian, Mississippian, Pennsylvanian, and Permian. All but the Permian are represented in Tennessee's geology, and they are found primarily in Middle and East Tennessee.

Table 4. Generalized Geologic Time Scale

Era	Period	Beginning of period (million years before present)	Dominant organisms in the life record	
			Animals	Plants
Cenozoic	Quarternary	2.5	Mammals	Flowering plants
	Tertiary	60		
Mesozoic	Cretaceous	140		
	Jurassic	205	Reptiles	Conifers, cycads
	Triassic	230		
Paleozoic	Permian	285		
	Pennsylvanian	325		Spore-bearing
	Mississippian	350		land plants
	Devonian	410		
	Silurian	430		
	Ordovician	500	Marine	Marine
	Cambrian	570	invertebrates	plants
Precambrian Eon	Proterozoic Z		One-celled organisms	
	Proterozoic Y	4,000+	Origin of Earth	

Following our time sequence, the next to youngest era of time is the Mesozoic ("middle life"). The Mesozoic era spanned a space of time beginning approximately 230 million years ago and ending about 65 million years ago. The Mesozoic era is divided into three periods of time (oldest to youngest): Triassic, Jurassic, and Cretaceous. Mesozoic rocks of Cretaceous age are found only in West Tennessee. Rocks representing the Triassic and Jurassic periods have not been found in Tennessee.

The most recent era, the one in which we live, is called the Cenozoic ("recent life"). The Cenozoic era, spanning a time interval from 65 million years ago to the present, is divided into two periods: the Tertiary is the older and the Quaternary the younger. Cenozoic age rocks and deposits are found over a large portion of West Tennessee, along most streams and rivers in the state (in the form of old alluvial deposits), and as colluvium, talus, and other recent unconsolidated surface deposits. In fact, the very landscape we see across Tennessee today was formed predominantly during Cenozoic time.

Fossils

A discussion of earth's history is not complete without a view of past life through fossils. Fossils are simply the remains or traces of plants or animals that have been preserved by natural processes in the earth's crust. Tennessee boasts a rich fossil record, with fossil animal and plant life found in all three major regions of the state (Fig. 18).

Recent studies have shown that the earliest life forms began as far back as 3.5 billion years ago (Knoll 1991: 67). The dominant life form throughout most of earth's history has been single-celled organisms.

Abundant forms of complex life began appearing in the fossil record some half billion years ago. A progressive and increasingly complex development of life is borne out by the fossil record. Earliest forms of life are thought to have been bacteria. Possibly emerging about the same time were single-celled plants and animals. Later, more complex forms of life emerged, with simple plants, invertebrates, and early forms of fish and amphibians dominating the Paleozoic era; dinosaurs the Mesozoic; and mammals, flowering plants, and insects the Cenozoic.

The important pattern to recognize in the fossil record is the increasing diversity and complexity of organisms through time. Complex forms of life such as the primates did not occur early in the history of the earth, but very late. Over hundreds of millions of years

Fig. 18. Brachiopods are commonly found in Paleozoic rocks of Tennessee.

the fragile entity of life has endured through adaptations, changes, and a never-ending drive to continue the species.

Fossils are found all across Tennessee. Microscopic forms include spores, pollen, and foraminifera. Macroscopic fossils (those that can be seen with the naked eye) include everything from worm burrows, brachiopods, trilobites (Fig. 19), corals, and sponges to pelecypods, crinoids, crabs, sharks' teeth, and even the bones of saber-toothed cats.

Fossils of vertebrate animals found in Tennessee rocks include sharks' teeth, mosasaur bones, and turtle remains of Cretaceous age. The teeth and bones of mastodons, saber-toothed cats, and even giant ground sloths (dating from the Pleistocene epoch, or Ice Age) have been found in caves, sinkholes, and old alluvial deposits.

Fossils are the direct or indirect remains of organisms preserved in the rock strata. Examples of direct evidence of life include actual remains, petrifactions, and prints. Indirect evidence of past life include molds or casts of actual remains, artifacts, trails, burrows, and tracks and coprolites (fossil excrements). To have fossils there must be conditions favorable to preservation. Two such conditions are the quick burial of organisms in a protective medium and the possession of hard parts, such as shells or hard skeletons.

Fig. 19. Trilobites, a three-lobed (head, body, tail) marine arthropod are considered prize fossil finds. This head piece is from an *Ollinellus* trilobite found in the Cambrian-age Rome Formation in Grainger County.

In your travels across Tennessee, you will have opportunities to observe fossil plants and animals. In some places the fossils will be numerous and plentiful. In other places patience and a good eye will be required for viewing them. If you collect specimens, then be sure you do so with the intent to gain information. Collecting fossils just for the sake of collecting reduces the amount of valuable information that a scientist may need to understand the geology of a particular area. Collecting only a very few specimens, however, will probably be fine—perhaps in some cases just the way to inspire a young future scientist.

Tennessee's Geologic History

From the mountains of the Blue Ridge to the flat bottomland of the Mississippi River Alluvial Floodplain lies a record of immense geologic history that spans over a billion years of time. The varied landscapes of Tennessee between the mountains and the plain—from parallel ridges and valleys to vast regions of karst and cedar glades—owe their existence to two principal geologic factors: the underlying bedrock and the weathering and erosion of the land surface.

The rocks found across Tennessee range in age from Precambrian granite and gneiss (some of which are over one billion years old) to recently deposited alluvial material in streams and rivers. In between are found Paleozoic limestone, sandstone, and shale 250 to 500 million years old; Mesozoic sand, silt, and marl over 60 million years old; Tertiary sand, silt, and clay several million to near 60 million years old; and Quaternary sand, gravel, boulders, and loess up to a million years old (Map 4).

These rock layers, some of which are flat-lying, others highly contorted and faulted, are found in creek bottoms, bluffs, caves, quarry pits and mines, and roadway cuts. Rock outcroppings can also be found in forests and pastures, riverbanks and lakeshores, farms and backyards.

A trip across Tennessee from east to west provides an easy means for following the chronological sequence of the state's geology. This sequence is important not simply because it is chronological but because older rocks provide a foundation for the later rock layers. Not only that, the younger rocks are often composed of fragments of older ones. The sandstone forming in the Gulf of Mexico at the mouth of the

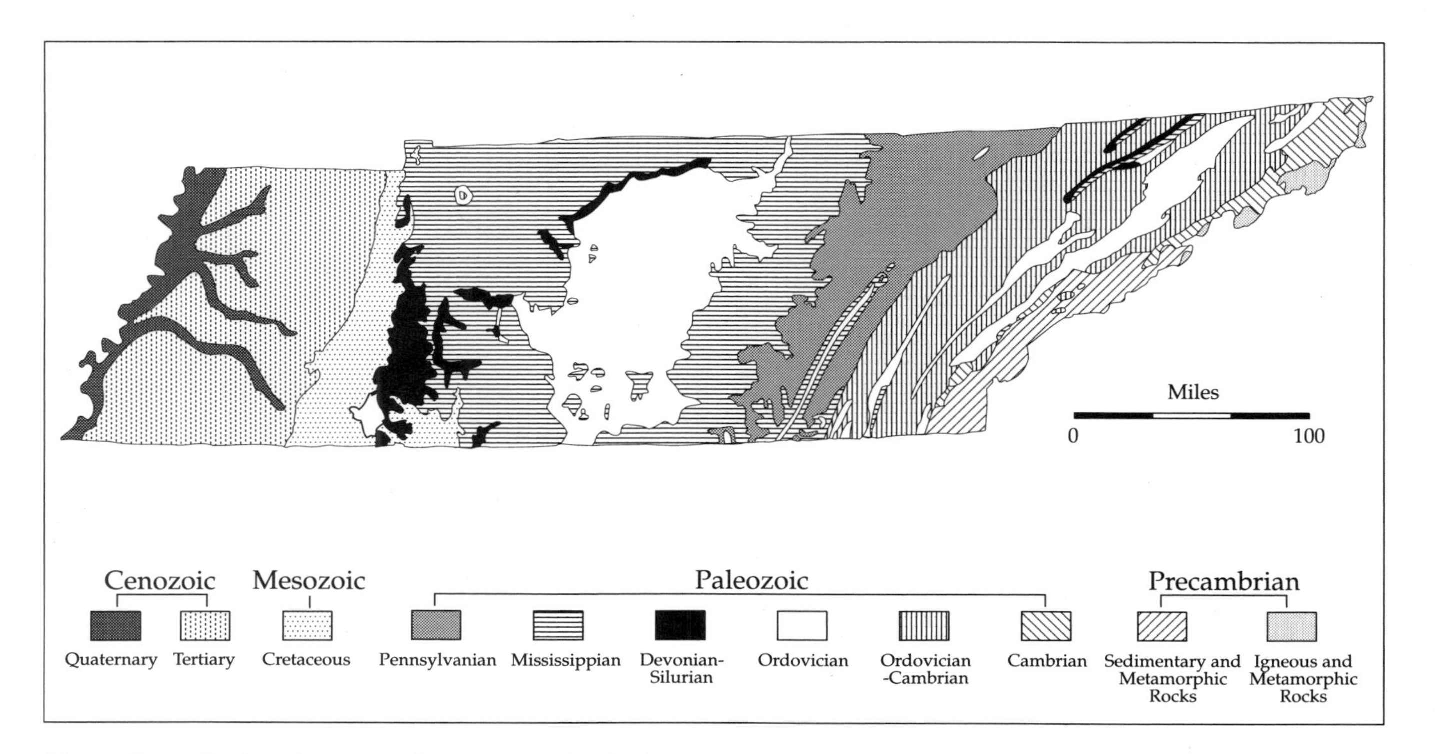

Map 4. Generalized geologic map of Tennessee. After Geologic Map of Tennessee, Tennessee Division of Geology, 1966.

Mississippi River, for example, originated as grains of quartz from sandstone of Cretaceous, or Pennsylvanian, or even Precambrian age, some of it far upstream from the new deposits.

On a geological journey across Tennessee, the physical travel from the Blue Ridge to the Mississippi River is paralleled by a journey in geological time from the Precambrian era, through the Paleozoic and Mesozoic eras, to the Tertiary and Quaternary periods of the current Cenozoic era. While the very oldest rocks show little evidence of early life, the gray, thin-bedded limestone of East and Middle Tennessee and the green marl and brown clay and sand of West Tennessee preserve a fascinating history of shallow seas and widespread marine life.

Although geologists, recognizing the complexity of the state's rocks, have given names to many individual formations, understanding the grand sweep of Tennessee geology requires distinguishing only a few major periods of time and some common types of rock (Table 5). Tables 6, 7, and 8 list the rock formations found in each of the three grand divisions of Tennessee: East (Table 6), Middle (Table 7), and West (Table 8).

Table 5. Age Distribution of Geologic Formations in Tennessee

Era	Period	Lithology	Region
Cenozoic	Quaternary	Alluvial sand and gravel	East, Middle, West
		Loess	West
Mesozoic	Tertiary	Clay, sand	West
	Cretaceous	Silt, sand, clay, marl, gravel	West, west part of Middle
Paleozoic	Pennsylvanian	Shale, sandstone, siltstone, coal	East
	Mississippian	Sandstone, limestone, shale	East, Middle
	Devonian	Shale	East, Middle, parts of West
		Shale, limestone	Middle, east part of West
	Silurian	Shale, sandstone	East
		Shale, limestone	Middle, east part of West
	Ordovician	Limestone, dolostone, shale, sandstone	East
		Limestone, shale	Middle, West
	Cambrian	Limestone, shale, sandstone, dolostone	East
		Dolostone, limestone	East
Precambrian		Metasandstone, slate, granite, gneiss	East

Table 6. Geologic Groups and Formations in East Tennessee

Era	Period	Group	Formation	Province
Cenozoic	Quaternary		Terraced stream deposits, colluvial deposits	Blue Ridge, Valley and Ridge, Cumberland Plateau
Paleozoic	Pennsylvanian	Crab Orchard Mountain	Rockcastle, Whitwell, Sewanee	Cumberland Plateau
		Gizzard Group	Signal Point, Warren Point, Raccoon Mountain	Cumberland Plateau
	Mississippian		Pennington	Valley and Ridge, Cumberland Plateau
			Fort Payne, Grainger	Valley and Ridge
	Devonian		Chattanooga	Valley and Ridge, Cumberland Plateau
	Silurian		Rockwood, Clinch	Valley and Ridge
	Ordovician	Chickamauga Group	Moccasin/Bays, Ottosee/Sevier, Chapman Ridge, Holston, Lenoir	Valley and Ridge
		Knox Group		Valley and Ridge
	Cambrian	Knox Group		Valley and Ridge, Blue Ridge
		Conasauga Group	Honaker	Valley and Ridge
			Rome, Shady	Valley and Ridge, Blue Ridge
		Chilhowee Group	Hesse, Murray, Nebo, Nichols	Valley and Ridge, Blue Ridge
Precambrian		Ocoee Supergroup		Blue Ridge
		Walden Creek Group	Wilhite, Cades	Blue Ridge
		Great Smoky Group	Elkmont	Blue Ridge
		Snowbird Group	Pigeon Siltstone, Roaring Fork Sandstone, Metcalf	Blue Ridge
		Mount Rogers Group		Blue Ridge
		Basement Complex	Beech Granite	Blue Ridge

Note: The table lists only those rock groups and formations mentioned in the text.

Table 7. Geologic Groups and Formations in Middle Tennessee

Era	Period	Group	Formation	Province
Cenozoic	Quaternary		Alluvial deposits, colluvial deposits	Eastern Highland Rim, Central Basin, Western Highland Rim
Mesozoic	Cretaceous		Tuscaloosa	Western Highland Rim, Gulf Coastal Plain
Paleozoic	Mississippian		Pennington	Eastern Highland Rim
			St. Louis, Warsaw, Fort Payne (Maury Shale)	Eastern Highland Rim, Western Highland Rim
	Devonian		Chattanooga	Eastern Highland Rim, Western Highland Rim
			Pegram, Ross/Birdsong Shale	Central Basin
	Silurian		Brownsport	Western Highland Rim, Western Valley of the Tennessee River
		Wayne Group	Waldron Shale, Brassfield	Western Highland Rim, Western Valley of the Tennessee River
	Ordovician		Liepers	Central Basin
		Nashville Group	Catheys, Bigby-Cannon, Hermitage	Central Basin
		Stones River Group	Carters, Lebanon, Ridley	Central Basin
	Cambrian	Knox Group		Western Highland Rim

Note: The table lists only those rock groups and formations mentioned in the text.

Table 8. Geologic Groups and Formations in West Tennessee

Era	Period	Group	Formation	Province
Cenozoic	Quaternary		Alluvial deposits	Western Valley of the Tennessee River, Gulf Coastal Plain, Mississippi River Alluvial Floodplain
			Loess	Gulf Coastal Plain
	Tertiary		Jackson, Claiborne, Wilcox	Gulf Coastal Plain
		Midway Group	Porters Creek, Clayton	Gulf Coastal Plain
Mesozoic	Cretaceous		McNairy, Coon Creek, Coffee	Gulf Coastal Plain
			Tuscaloosa	Western Valley of the Tennessee River, Gulf Coastal Plain
Paleozoic	Devonian		Chattanooga, Pegram,	Western Valley of the Tennessee River, Gulf Coastal Plain
			Ross/Birdsong Shale	Western Valley of the Tennessee River, Gulf Coastal Plain
	Silurian	Wayne Group	Brownsport, Waldron Shale	Western Valley of the Tennessee River
			Brassfield	Gulf Coastal Plain

Note: The table lists only those rock groups and formations mentioned in the text.

Precambrian Eon

Rocks of Precambrian age are exposed in the mountainous regions of East Tennessee in the Blue Ridge province. Exposures of Precambrian rocks can be found in Polk, Monroe, Blount, Sevier, Cocke, Greene, Unicoi, Carter, and Johnson counties.

Most of the Precambrian rocks are either granitic or sedimentary in origin. Because metamorphism has greatly altered some of the rock layers, however, their true origin is often hard to determine. Most of the Precambrian sedimentary rocks have been deformed and affected by metamorphism. Geologists have divided the Precambrian rocks in Tennessee into the older basement complex and the younger Ocoee Supergroup. These subdivisions, based upon lithologic (rock) types and mappable units, are further subdivided into groups and formations (King et al. 1968: 2).

The oldest of the Precambrian rocks are believed to be mostly igneous in origin (some possibly sedimentary) and are represented by granite, gneiss, schist, and gabbro rock types (Fig. 20). These rocks, which comprise the basement complex, are found in the northeasternmost sections of East Tennessee in Cocke, Unicoi, Carter, and Johnson counties. Uranium-lead dating has shown that some of the rocks, known as the Cranberry Granite, are at least one billion years old (Miller 1974: 14).

Overlying the older basement complex is the Ocoee Supergroup, a sequence of slightly to moderately metamorphosed sedimentary rocks. The Ocoee Supergroup includes four groups of formations. These altered sedimentary rocks consist mainly of quartzite, slate, phyllite, and schist that have been folded, fractured, and faulted into a complex arrangement. The sedimentary equivalent of these are sandstone, siltstone, and shale. Metamorphosed limestone is called marble.

The Ocoee Supergroup is well exposed in the Great Smoky Mountains National Park, where about 80 percent of the rocks belong to the Ocoee Supergroup (King 1968: 3; Moore 1988: 31). In the southeast corner of Tennessee in Polk County, the rocks of the Ocoee Supergroup have been significantly altered through metamorphism to produce highly mineralized zones. These mineral deposits (mostly copper-iron sulfides) were mined in the Ducktown-Copperhill area to obtain sulfuric acid, as well as smaller quantities of copper and gold, from the sulfide-rich rocks.

Fig. 20. Some of the oldest rocks in Tennessee are the Precambrian gneisses found in Unicoi County, some of which are over a billion years old.

Most of the Precambrian rocks are unfossiliferous: they do not contain fossils, for they represent a time period when primeval life was just developing. Life forms found in Precambrian rocks elsewhere in North America are primitive, consisting mainly of algae and soft-bodied animals such as jellyfish. Recent reports of an invertebrate fossil in the upper part of the Ocoee Supergroup–Walden Creek Group in Monroe and Polk counties have been the subject of great interest, as invertebrate fossils have not often been found in these Precambrian rocks (Unrug and Unrug 1990: 1041–45).

Recent highway construction in Unicoi County has exposed units of the Precambrian basement complex. Rock types such as granite gneiss, augen gneiss, granite, graphitic phyllite, and greenstone are exposed in the numerous new road cutslopes. One variety of the granitic material exposed in Unicoi County is known as unakite, a pink-and-green rock that is greatly sought after by many rockhounds.

Paleozoic Era

During the time period from approximately 570 million to 230 million years ago, the earth's continental and oceanic plates were on the move, shifting seas and oceans, building mountains, and accumulating large thicknesses of sedimentary rock. Life forms, primitively developed in late Precambrian time, exploded into a profusion of plants and animals that marked an accelerated path of evolution. During this evolutionary surge, life made its transition from the seas and oceans to terrestrial habitats. This 340-million-year time span is referred to by geologists as the Paleozoic era. Six of the seven geologic periods of the Paleozoic era are represented in the rock units in Tennessee. Only the rocks of Permian age are absent.

Cambrian Period

The Cambrian rocks of Tennessee consist of shale, siltstone, sandstone, quartzite, dolomite, and limestone. Most of the Cambrian age rocks are found in East Tennessee in the Valley and Ridge and the Blue Ridge provinces. Although some Cambrian age limestone is found in Middle Tennessee, the exposures are limited; noteworthy are the Wells Creek, Howell, and Flynn Creek structures (see Side Trip 7 to Wells Creek structure and Dunbar Cave State Natural Area).

The oldest of the Cambrian age rocks are represented by strata referred to as the Chilhowee Group, whose formations include quartzite, sandstone, siltstone, and shale. The area in which strata of the Chilhowee Group are exposed is confined to the eastern portions of East Tennessee. Due to the resistant nature of the quartzite and sandstone found in the Chilhowee Group, a number of large northeast-southwest trending mountains are formed by their lithologies. Most notable of these mountains are Starr Mountain (Polk County), Chilhowee Mountain (Blount and Sevier counties), English Mountain (Sevier and Cocke counties), Holston Mountain (Sullivan and Johnson counties), and Iron Mountain (Carter and Johnson counties).

The rocks of the Chilhowee Group represent an environment in which tremendous quantities of sediment were being eroded from a landmass probably located to the west of the area where the rocks are presently exposed (Miller 1974: 16). Recent geological research has revealed that paleocurrent data (compass readings of current features, such as ripple marks and cross-bedding, preserved in the rock layers) support such a westward source of the sediment (J. D. Walker 1990: 9). Large influxes of eroded debris were deposited into a sea basin—a relatively shallow one, as the presence of ripple marks, cross-bedding, possible worm trails and burrows, and some fossil ostracods indicates. (Some of the largest "fossilized" ripple marks ever found in Tennessee occur in strata of the Chilhowee Group south of Erwin in Unicoi County.)

At the end of the deposition of the Chilhowee sediment, a major change in the composition of the sedimentary rocks began as a result of modification of the ocean environment—a shallow bank. The introduction of major units of carbonate rocks (limestone and dolostone) into the geologic column is initially represented by the Shady Formation in Tennessee (Miller 1974: 16). Although some units of the older late Precambrian Wilhite Formation (Tellico Mountains area) contain some carbonate horizons, they are not classified as a separate formation.

The remaining Cambrian time is represented predominantly by limestone, dolostone, and shale lithologies; the carbonate units are more widespread in the northeastern part of East Tennessee. After the Shady Formation was deposited, a more clastic sequence (that is, composed of sand, silt, and clay derived from the weathering of other rocks) accumulated as sediments in shallow water to create the Rome Formation. Noted for its variegated shale and siltstone, the Rome

Formation is also associated with a number of large-scale thrust faults, including the Saltville and Copper Creek faults (near Knoxville).

A sequence of shale and limestone called the Conasauga Group overlies the Rome Formation. The strata of the Conasauga Group gradually change northeastward into the Honaker Formation, a medium- to thick-bedded dolostone. From the transition between the Cambrian period and the Ordovician period comes a sequence of carbonate strata known as the Knox Group.

As mentioned earlier, the shallow seas to the east were being supplied with sediment from an eroding craton to the west, developing a broad continental shelf environment where large sequences of limestone, shaly limestone, and dolostone formed.

Present in this shallow sea were numerous life forms that flourished in the Cambrian era and continued their evolution throughout the Paleozoic era. The earliest of these is known as *scolithus,* the fossil of what was probably a worm boring (King et al. 1968, 10; Miller 1974: 18). These fossil borings appear as dark, nearly vertical lines, 6 to 10 inches long, throughout the strata of the Chilhowee Group. The borings are particularly abundant in the Hesse and Nichols formations, where they are usually pencil-sized.

Also found in the Chilhowee Group are ostracods, small bivalve crustaceans that tend to live in brackish, shallow marine waters. These fossils, found in the Murray Shale on Chilhowee Mountain (Lawrence 1963: 53–54), mark the earliest occurrence in Tennessee of discernible fossilized animal parts.

Other notable fossils found in Cambrian rocks include trilobites and brachiopods. Trilobites, marine arthropods whose bodies show three distinct longitudinal lobes, probably scavenged along the ocean shores, like their descendants, the horseshoe crabs. Brachiopods are a phylum of marine animals with two unequal shells, which get their name (meaning "arm-foot") from the appearance of the appendages that sweep food into their mouths, appendages that early scholars mistook for limbs.

The trilobites are so abundant in Cambrian rocks that geologists have nicknamed the period the "Age of Trilobites." Most notable among the Cambrian trilobites occurring in Tennessee is the genus *Olenellus,* which is used worldwide as a Cambrian index fossil. That is to say, a species (such as *Olenellus*) that is common in a particular type of rock but apparently absent from others, can be used as a diagnostic tool to identify the age of those rocks. In Tennessee it is found in the lower to middle Cambrian age Rome Formation. Several other Cambrian trilobites are also found in the strata of the Conasauga Group (McLaughlin 1973: 28–30).

In the limestone and dolostone of the Knox Group are assemblages of cabbage- to watermelon-sized masses of thinly laminated deposits. These round masses, organic in origin, are composed predominantly of algal material called stromatolites. Present-day examples can be found along the west coast of Australia.

Ordovician Period

In Tennessee the Ordovician period is marked by the deposition of thick sequences of carbonates and shale. These rock units are found throughout much of East and Middle Tennessee, primarily in the Valley and Ridge and Central Basin areas.

The earliest of the Ordovician strata are the carbonates of the Knox Group, which began accumulating in uppermost Cambrian time. The depositional environment of the Knox Group was typically supratidal to intertidal mud flats (above high tide to low tide) of carbonate composition. When Knox carbonates were exposed above the existing sea in early Ordovician time, erosion produced a karst surface full of caves, sinkholes, and depressions. Some of the ceilings of these caves apparently collapsed, forming extensive breccia deposits (angular rock fragments) in the upper portions of the Knox Group. These breccia deposits have since mineralized, forming one of the richest zinc deposits in the U.S., known as the Mascot–Jefferson City Mining District in East Tennessee. These "fossil" sinkholes and depressions, called paleokarst, are visible in sections of Douglas Lake in Sevier and Jefferson counties during periods of low water levels.

As the seas advanced back over the paleokarst land surface, deposition of marine sediments into the exposed sinkholes and depressions began, marking the beginning of middle Ordovician time. Rocks formed from these sediments are referred to as the Chickamauga Group in East Tennessee and the Stones River and Nashville groups in Middle Tennessee.

Of particular interest in the Chickamauga Group is the development of bryozoan reef environments. These reefs, complex accumulations of living organisms, consisted of front reef, reef core, and back reef environments (Walker and Ferrigno 1973: 131). Today, these rocks are referred to as the Holston Formation, which is found principally in the middle portions of the Valley and Ridge area (Knoxville vicinity). The

Holston strata were extensively quarried during the early half of the twentieth century for their pink and gray marble-like stone. Although not a true marble, the limestone is coarse crystalline and polishes well; it is marketed nationally as "Tennessee Marble." A number of buildings in the Smithsonian complex in Washington, D.C. (including the new East Wing of the National Gallery) are constructed and faced with stone from the Holston Formation.

Westward of these reefs the shallow marine waters provided environments for numerous shelled marine animals. In Middle Tennessee these marine carbonate sediments (the Stones River Group and the Nashville Group) contain numerous fossils of these Ordovician animals (C. W. Wilson 1948: 16–18; Miller 1974: 20). During middle Ordovician time a large influx of clastic sediments (the Sevier Formation) marked a change in the tectonic regime that would not be reversed for the remainder of the Paleozoic time (Hatcher and Lemiszki 1991: 1).

The Ordovician rocks provide a wealth of economic resources for Tennessee. These include not only the "marble" industry, but also important zinc deposits in both East and Middle Tennessee. In addition, the Ordovician limestone and dolostone supply most of the state's need for crushed stone, cement, and agricultural lime. In Middle Tennessee these carbonate rocks (Bigby and Leipers Limestone) were once mined for their phosphate content (Miller 1974: 20). Present market conditions, however, have slowed the phosphate mining industry.

Mountain-building activity during the Ordovician period, referred to as the Taconic Orogeny, was centered northeast of Tennessee (Miller 1974: 20) but produced some effects in what is now East Tennessee. Some of the structures found in the rocks of the Great Smoky Mountains are believed to be the result of this orogeny. In addition, volcanic activity to the east (in what is now Virginia and North Carolina) spread ash deposits across large portions of the lands to the west, including Tennessee. These ash deposits settled into the sea and formed distinctive beds of sediment material that has since been chemically altered into bentonite clay (Miller 1974: 20). These bentonite clay layers are found throughout the Middle and Upper Ordovician rocks in the Central Basin and Valley and Ridge provinces (Miller 1974: 20).

Life in the Ordovician period was characterized by a growth in diversity and high population densities of marine invertebrates, particularly bryozoans. These are tiny colonial marine animals that build calcareous structures, commonly called reefs, primarily in the ocean but occasionally in fresh water. Bryozoans, whose name means "moss animals" in Greek, get the name from the resemblance of their colonies to collective plant growths like mosses, liverworts, and lichens. These reefs in the shallow Ordovician seas provided a habitat for numerous other invertebrates, such as the algae that were important contributors to the Ordovician ecosystem.

In addition to the bryozoans and algaes, numerous shelled invertebrates also occupied these warm shallow seas. Brachiopods can be found in profusion in numerous Ordovician limestone beds (McLaughlin 1973: 36–42). Marine snails (gastropods), which were also common, are found in the Lenoir Formation around the Knoxville area.

Corals, also abundant in the Ordovician seas, are commonly found fossilized in the Ordovician rocks of the Central Basin area (Miller 1974: 22). Other notable invertebrate phyla preserved in the limestone and shale include cephalopods, graptolites, ostracods, and trilobites. Cephalopods, so named because their limbs appear to grow directly out of their heads, today are represented by the nautilus, squid, and octopus. Most of them swim by ejecting a jet of water from under their mantle. Shelled cephalopods, the ancestors of the nautilus, seem to have been the most numerous form in the Ordovician period. The graptolites, which do not seem to have left any modern descendants, were colonial animals, with shells of chitin, or horn; they were so prevalent during the Ordovician period that it is sometimes called the Age of Graptolites. Ordovician trilobites can often be found in the Ottosee Shale of the Valley and Ridge province. Although the oldest fish fossils date from the Ordovician period, no such fish remains (ostracoderms) have been found in the Ordovician rocks of Tennessee.

Silurian Period

The Silurian period in Tennessee is represented by two contrasting marine environments. To the west, in the vicinity of the Western Highland Rim (west of Nashville) and the Western Valley of the Tennessee River, are strata that were formed in shallow seas and are profusely fossiliferous. These rocks are primarily thin-bedded limestone but are commonly interbedded with shale (muddy sediments washed into the shallow marine waters).

In contrast, to the east in the Valley and Ridge areas along the eastern Cumberland Escarpment and on Clinch and Powell mountains are sediments of mostly clastic origin. Sandstone, siltstone, and shale are the dominant Silurian lithologies found in East Tennessee. The Valley and Ridge Clinch Sandstone dominantly exposed along Clinch Mountain is interpreted as being a beach deposit of a westward advancing sea (Miller 1974: 23), which contrasts with the quiet shallow-water carbonate environment to the west.

A notable point about the rocks of the Silurian period is the influence they had on the history of East Tennessee. Found in the shale, siltstone, and sandstone units of the Rockwood Formation of East Tennessee are beds of hematite (red iron ore). These iron ore deposits were mined by the early settlers, especially along the eastern Cumberland Escarpment. The production of iron was crucial to the economic growth of East Tennessee. Communities such as Rockwood, LaFollette, Cumberland Gap, Dayton, and Chattanooga (initially called Ross Landing) all experienced growth because of mining and processing operations (Miller 1974: 23). Repercussions, however, included not only smog and soot-filled air but finally a "bust" in the iron industry.

Life during the Silurian period continued to evolve into many diverse forms. The Silurian rocks in Tennessee reflect this rapid growth and spread of marine life forms. Most groups of invertebrate animals are represented, including brachiopods, cephalopods, corals, crinoids, gastropods, sponges, and trilobites. Crinoids were marine animals of the echinoderm ("spiny-skin") phylum that includes starfish and sand dollars. Many crinoids were sessile (that is, they stayed in one place, anchored to the bottom), and the fossil remains most often found of crinoids are the long stems that supported the head, with its numerous radiating arms.

Again, the fossil assemblage found in the strata reflects the two differing environments in which the rocks were formed. In the east the Clinch Sandstone hosts an abundance of specimens of the Silurian index fossil *Arthrophycus* (Schoner 1985: 103). These fossils are actually casts of horizontal borings made in the sandy sediment of the beach sand environment by what was probably a polychaete marine worm. Commonly segmented, these fossil borings radiate horizontally 6 to 10 inches from a central beginning point and often overlap each other. Also found in the Clinch Sandstone is the fossil called *Scolithus:* vertical borings, 8 to 10 inches in length, thought to be those of a worm or crustacean. Both of these fossils are easily found on Clinch Mountain in Grainger County where highway construction has exposed numerous beds of the Clinch Formation (see Side Trip 1 to Cumberland Gap National Historic Park).

In the west part of Middle Tennessee is found a very different but profuse fossiliferous assemblage. The Waldron Shale, a 3-to-5-foot-thick greenish-gray rock unit exposed near Pegram, Tennessee, just west of Nashville (see Road Log, Miles 193.3 and 192.9) contains well-preserved fossils of a living environment replete with masses of corals, beds of crinoids, communities of bivalves, bryozoans, and gastropods. Numerous trilobites that scavenged the sea floor are excellently preserved in the shale strata.

Further west, in the Western Valley, is exposed a shaly limestone known as the Brownsport Formation. Excellent Silurian fauna are also found here, with fossil representatives of crinoids, corals, and sponges very numerous and easily collected along U.S. 641 and S.R. 69 (see Road Log, Mile 126.1). Other fossil representatives that can be found in Silurian age rocks of Tennessee include brachiopods, sponges, crinoids, and corals (Miller 1974: 24–25).

Devonian Period

Devonian age rocks are found in both Middle and East Tennessee. In Middle Tennessee they are found mainly along the Western Valley of the Tennessee River and along the marginal boundaries of the Highland Rim around the Central Basin. In East Tennessee, Devonian strata are found sporadically across the Valley and Ridge but occur mainly along Clinch and Powell mountains, the lower portions of Chilhowee Mountain (Blount County), near Chattanooga, and along the boundaries of Sequatchie Valley.

During the Devonian period, two distinct environments of deposition were present: a shallow marine environment, in which numerous shelled organisms lived, and a quiet water environment, in which the deposition of a black mud was most widespread. Early in Devonian time the shallow seas covered the Western Valley area, where a fossiliferous calcareous mud sediment was deposited. At the same time in the Valley and Ridge, the carbonate sediments of late Silurian time continued accumulating into early Devonian time.

Near the end of early Devonian time, the land was uplifted and subject to erosion. The seas advanced over the eroded land surface, depositing sediment now represented by the Pegram Formation. Such a boundary between two formations, delineated by an erosional surface, is called an unconformity.

After deposition of the Pegram Formation during Middle Devonian time, the seas retreated and subjected the land surface to renewed erosion activity. Again the erosion was extensive, in some places removing rocks of Silurian and Ordovician age as well as those of early Devonian age (Miller 1974: 26).

During late Devonian time the seas again advanced over this eroded land surface. A black mud, rich in organic matter, began to be deposited. Eventually, deposition of the carbonaceous mud sediments, now called the Chattanooga Shale, extended over much of the East-Central United States, including Tennessee.

In Middle Tennessee the Chattanooga Shale is relatively thin (about 20 feet) but thickens eastward to well over 100 feet. On outcrop it is a very thin, fissile shale easily recognizable by its distinctive black color (see Road Log, Mile 271.7).

The Chattanooga Shale is typical of the sedimentary rock that forms where once there was a swamp or lagoon (or perhaps a deep ocean basin). Because the waters that produced these sediments were calm, the organisms in them consumed most of the oxygen in the water. In such conditions, called a reducing environment, minerals such as pyrite (iron sulfide) may form and petroleum may develop from the hydrocarbons in the organic matter. The Chattanooga Shale is relatively rich in petroleum (15 gallons per ton) and contains a trace of uranium (Miller 1974: 28).

Life during the Devonian period was at first profuse and varied. The Devonian period is often called the Age of Fishes because of the numerous fish fossils found in Devonian age rocks. Although fish fossils are not plentiful in Tennessee Devonian strata, a few pieces have been found, marking the first occurrence of vertebrate remains in Tennessee rocks (Miller 1974: 28).

During early Devonian time in Tennessee, many invertebrates inhabited the shallow seas. Among these invertebrates were such animal groups as brachiopods, crinoids, trilobites, and conodonts (Dunbar 1919: 34–93). What animal produced the fossils known as conodonts is still a matter of speculation, but they are recognized by their appearance as conical, tooth-like structures of calcium phosphate. In the Birdsong Shale member of the Ross Formation, found in the Western Valley area (see Road Log, Mile 133.0), a varied accumulation of invertebrates can be found. The most abundant invertebrates were the brachiopods. Another notable invertebrate was the trilobite *Dalmanites,* a distinctive species that has been found up to 12 inches long in Tennessee (Miller 1974: 28).

Late in Devonian time, life was significantly different, but not many preserved examples are to be found in Tennessee. Though scarce, fossils occurring in the Chattanooga Shale include lingulid brachiopods (a family of brachiopods with simple shells) and some conodonts. Some particularly well-preserved lingulid brachiopods have been found near the base of Clinch Mountain (see Side Trip 1 to Cumberland Gap National Historic Park). The deposition of the Chattanooga sediments continued into early Mississippian time.

Mississippian Period

Mississippian time in Tennessee is characterized by shallow seas and limey sediment (Fig. 21). In these seas lived a varied and prolific life that adjusted and evolved in response to the changing Mississippian seas. Early Mississippian time (Lower Mississippian) is marked by the deposition of a muddy sediment called the Maury Shale, which covered most areas of Tennessee. This formation, one to two feet thick, is easily distinguished as a light greenish-gray shale that in most places lies directly on the black Chattanooga Shale (see Road Log, Mile 271.7). Some researchers have reported the presence of fish bone fragments in the shale material (Miller 1974: 31).

Succeeding the deposition of the Maury Shale was the deposition of the Fort Payne Formation, a silty calcareous sediment. The Fort Payne Formation, largely replaced by silica, is now a very silicious, cherty-bedded accumulation of limestone, dolostone, and thin, silty shale. Large accumulations of marine organisms in reef-like structures characterize the Fort Payne in the Eastern Highland Rim/Cumberland Plateau area (Miller 1974: 31). Strata of the Fort Payne Formation are easily identified as beds of yellowish-gray, silicious limestone, one to two feet thick, along I-40, near the edge of the Eastern Highland Rim (Mile 271.7).

In Middle Tennessee most of the Mississippian rocks deposited after Fort Payne time (Upper Mississippian) are limestone. Collectively they represent a time of shallow warm seas with strong currents, as exemplified by cross-bedded limestone. In East Tennessee the Upper Mississippian rocks are characterized by a mix of clastic and calcareous sediments.

The Upper Mississippian strata are capped by the Pennington Formation, an accumulation of mostly clastic sediments, including shale, siltstone, and sandstone. Outcroppings of the Pennington Formation, characterized by maroon and greenish-gray shale, can be

Fig. 21. In Middle Tennessee the layers of rock are horizontal, much like the position in which they were formed.

found in both Middle and East Tennessee. A peculiar aspect of the Pennington Formation is its association with many areas of slope instability along the Cumberland Escarpment. Highways constructed across this formation along the escarpment have accordingly experienced numerous embankment failures.

The flourishing life during Mississippian time was dominated by invertebrate marine organisms. The Mississippian period is often called the Age of Crinoids because the fossil record indicates that crinoids seem to have diversified and prolifically populated the warm Mississippian seas. Much Mississippian age limestone in Tennessee bears the fossil remains of crinoids. In some places such large colonies of crinoids populated certain marine environments that they formed great mounds of crinoid debris when they died and fell to the sea bottom (Miller 1974: 31). This was the case with the Fort Payne Formation (Road Log, Mile 271.7).

Other notable invertebrates of Mississippian time were corals, which included both solitary horn corals (shaped like a small cow horn) and beautifully preserved colony corals (which grew in large masses, not unlike the corals of today). In addition, brachiopods and fan-like bryozoans ("moss animals") also populated the Mississippian waters.

A small, one-celled invertebrate organism known as *foraminifera* lived in the warm sea waters and populated to great numbers during the Mississippian period in Tennessee. The foraminifera had very small calcium carbonate shells that accumulated as a limey mud on the sea floor. Much of the Mississippian age limestone in Tennessee, as well as in other states, contains beds of these foraminifera shells.

Pennsylvanian Period

As Mississippian time came to a close and Pennsylvanian time began, a great change in the deposition of sediment occurred. The North American and African continents continued to move closer together because of continental drift, the movement of oceanic and continental plates. As a result, the marine environment also changed, reflecting changes in erosion and sedimentation rates as well as changes in the living organisms of that time period.

In Tennessee the Pennsylvanian age rocks reflect this change by being dominantly sandstone, siltstone, and shale (clastic sediments) rather than the carbonate rocks of earlier Paleozoic time. In addition, the presence of coal in Pennsylvanian age rocks is significant in the geologic history of not only Tennessee but of the world for it signifies the emergence of land plants on the continents. The land plants were so plentiful in certain environments (swamps, for example) that their decaying debris accumulated in such thick quantities that the result was the formation of layers of coal.

In Tennessee the Pennsylvanian age rocks are found principally on the Cumberland Plateau, where they form the caprock (the hard, resistant rock that erosion has failed to weather away). About 20 miles to the west of the plateau, Short Mountain, near Woodbury in Cannon County, is also capped by Pennsylvanian sediments, which probably extended west across the Highland Rim and the Central Basin areas (Miller 1974: 32). Erosion has removed these Pennsylvanian rocks, leaving only Short Mountain, a monadnock, as evidence of a once greater extent across Tennessee. A monadnock is an erosional remnant of older topography; it rises all alone above the present plain of subaerial erosion. East of the plateau Pennsylvanian age rocks are found only on Grindstone Mountain near Ooltewah in Hamilton County.

Along the coast of an ocean are various zones defined by the interaction of land and water, known as environments of deposition. These environments include the subtidal areas below the lowest tide line, the intertidal areas between high- and low-tide marks, such as the typical beach (Fig. 22), and the coastal margin, the area above the high-tide line that may include estuaries, lagoons, and swamps. In the Pennsylvanian age these environments were moving—landward, then seaward, and back again. The dominant environment seemed to be a complex of coastal barrier islands, tidal flats, lagoons, swamps, and typical intertidal sandy beaches (Ferm, Milici, and Eason 1972: 4–7). During Pennsylvanian time, these constantly moving areas (some of which were prograding deltas, advancing into the sea from the coastline) left a sequence of rock units consisting mostly of sandstone, siltstone, shale, and coal.

As these mostly clastic sediments accumulated, different rocks were formed, "genetically" tied to their environment of deposition. The great sandy beach barrier islands and migrating river channels were preserved as thick, well-indurated, cross-bedded sandstone and conglomerate. (An indurated rock is one that has become hard by cementation as well as by being exposed to heat or pressure.) The mud and organic debris that accumulated in the landward lagoon and tidal flats were preserved as the

Fig. 22. Ripple marks preserved in rock strata attest to their origin in shallow water, much like the coast of modern-day landmasses.

thick accumulations of gray and black shale. Organic debris from the numerous swamps resulted in the formation of coal seams. As the beaches and deltas migrated landward, sand was deposited over the gray mud of the lagoons and organic deposits of the swamps.

The types and abundance of vegetation found in Pennsylvanian deposits seem to indicate that this was a time of tropical swamps in Tennessee. Giant ferns and rushes grew, and died, to form great mats of decaying organic matter that were buried under the advancing and retreating sand of the beaches. Now these vast layers of organic matter appear as the bituminous coal beds of the Cumberland Plateau.

An unusual feature preserved in the Pennsylvanian rocks is the effect of channel-scour cutting through the barrier sand of the beaches along the coast. Coastal creeks and rivers, as well as streams draining lagoons and tidal flats behind the beaches, cut through the beach fronts to deposit sand and conglomerate in wedge-shaped deposits that differ from the surrounding mud and sand. An example of such a channel deposit can be clearly seen along I-40 at Mile 298.8 (Fig. 23). Early geologists interpreted this structure as an unconformity, an erosional surface commonly placed along the Mississippian/Pennsylvanian boundary. Researchers in the past twenty years have redefined these structures as coastal channel deposits (Ferm, Milici, and Eason 1972: 6).

Fig. 23. On the Cumberland Plateau, Pennsylvanian-age strata were formed in an environment much like that of the southeastern United States coast today.

The complexity of the coastal environments during the Pennsylvanian period is affirmed by the numerous intertonguing facies and depositional sequences of shale, sandstone, and coal. Intertonguing facies are layers of rocks that intrude or grade into each other, changing from one kind to another and back again, perhaps from shale to sandstone and back to shale (somewhat analogous to a double cheeseburger). The beach barrier sand, for example, can be found as the Warren Point and Sewanee formations at Fall Creek Falls and along the rim of Sequatchie Valley (R. L. Wilson 1981: 11) (see Side Trip 4 to Fall Creek Falls State Park). The shale of the Whitwell Formation represents the vast mud of back beach lagoons and swamps (Miller 1974: 32). Organic deposits of the swamps became coal deposits such as the Sewanee Coal of the Whitwell Formation.

Life during the Pennsylvanian period can be characterized as predominantly terrestrial swamp forest and coastal in nature. Accordingly, it is often called the Age of Forests. Huge "scale trees" and ferns grew in these swamp forests, with some growing to over 5 feet in diameter. The scale trees are identified as *Lepidodendron,* whose bark resembles a network of fingernail-sized plates or scales, such as fish scales. Large rushes known as *Calamites* grew up to 30 feet in height and 1 foot in diameter. (In comparison, present-day bulrushes found in Tennessee have an average height of 18 inches and a diameter of 1/4 inch.) Other plants, such as the cane-like *Sigillaria,* grew among the rushes and ferns. The fossil *Stigmaria* is believed to be the fossilized root system of the much larger plants.

Other life forms are represented by a few rare bivalve species. The first reptile fossils are found in the Pennsylvanian rocks of other states, but fish scales are the only vertebrate remains present in Tennessee rocks of Pennsylvanian age (Miller 1974: 34).

The last period of the Paleozoic era, the Permian, is not known to be preserved in Tennessee. The nearest Permian age rocks are in West Virginia and Texas. Perhaps erosion removed these rocks or Tennessee was above sea level during this time.

As the Paleozoic era drew to a close, a tremendous change in the North American continent took place. This change drastically affected the rocks of Tennessee and surrounding states. For 250 million years during the Paleozoic era and extending millions of years back into the Precambrian, sediments were accumulating along the east coast of the North American continent. During this time the continents of Africa and North America and the ocean plates in between were in motion, gradually closing in on each other. At the close of Paleozoic time, these sediments were gradually buckled, fractured, folded, and faulted as a result of the collision of the continents. This orogenic episode, the last to affect the Southern Appalachians, is referred to as the Alleghanian Orogeny.

The Allegheny mountain-building episode deformed the Appalachian rocks into their present structure. The rocks of the Piedmont, Blue Ridge, Valley and Ridge, and Plateau provinces from Pennsylvania and New York to Northern Alabama and Georgia were affected by this orogeny.

After the Alleghanian Orogeny much of the interior of eastern North America was above sea level and has remained so to the present. Notable exceptions are the Gulf Coastal Plain along the present East Coast and Gulf Coast and the Mississippi Embayment, including a portion of West Tennessee, which were again inundated by the sea.

At the close of the Alleghanian Orogeny, Tennessee was subject to extensive erosion, which lasted into Cretaceous time in West Tennessee and to the present in East Tennessee. The close of the Paleozoic era is marked by the deformation of the rocks in East Tennessee and the onset of millions of years of erosion.

Mesozoic Era

Commonly called the Age of Reptiles because dinosaurs ruled the earth then, the Mesozoic era represents a time of renewed diversity of organisms and prodigious evolution. The Mesozoic era is divided into three time periods: Triassic, Jurassic, and Cretaceous (oldest to youngest). In Tennessee only strata of the Cretaceous period are known to be represented. However, the erosional effects during the Triassic and Jurassic periods were reflected on the Tennessee landscape.

As the land area continued to rise in the east, the seas retreated to the west. During the Triassic period the area now occupied by Sequatchie Valley was but a low linear ridge or mountain quite possibly similar to the present-day Crab Orchard Mountain in Cumberland County (Road Log, Mile 326.0).

Rivers that developed as the land continued to rise above sea level began eroding the late Paleozoic sediments in Tennessee. Researchers theorize that the Tennessee River Gorge through Walden Ridge, just west of Chattanooga, was initially cut by the early Triassic drainage (Milici 1968: 191; Miller 1974: 42). Erosion also began the cutting of Sequatchie Valley.

By Jurassic time the erosion of the Tennessee landscape was well underway. Erosion had breached the higher portions of the Nashville dome, initiating the development of the Central Basin landscape. The Pennsylvanian sandstone and shale were being eroded off most of the Middle and West Tennessee area.

By Cretaceous time the Tennessee landscape was further developed. The Central Basin was appreciably excavated, with escarpments in most directions but principally to the east. Erosion had stripped away the Pennsylvanian strata and penetrated the Mississippian limestone, producing a karst landscape full of sinkholes (Miller 1974: 44).

During the Triassic and Jurassic periods, enormous amounts of rock were eroded from the surface by well-established drainage networks. The major topographic regions of the state—the Blue Ridge, Valley and Ridge, Cumberland Plateau, Highland Rim, and Central Basin—though not at their present configurations and elevations by Cretaceous time, were already significantly developed.

Cretaceous Period

The Cretaceous period marks a time of prolific evolution of the plant and animal species of the world. It was during Cretaceous time that the dinosaurs reached their maximum diversity and dominated the earth. The period is divided into three sections of time: early, middle, and late. In Tennessee only sediments of the late Cretaceous period are preserved, and these are found primarily in West Tennessee. Extensive research on the stratigraphy of the Cretaceous sediments of Tennessee has revealed an intriguing history of migrating marine environments and subsequent deposits of marine sediments (lithofacies) that represents transgression and regression of the Cretaceous sea (Russell 1965; Russell and Parks 1975; Stearns 1958; Marcher and Stearns 1962; Stearns and Armstrong 1955; and Stearns and Marcher 1962).

It is thought that a highland of Paleozoic rocks as old as Cambrian age, called the Clifton Saddle, existed in the West Tennessee region now occupied by Dyer, Lake, and Obion counties (Stearns and Marcher 1962: 1393). This "saddle" was actually an extension of the Ozark Dome that connected with the western flank of the Nashville dome.

Erosion of this highland area deposited cherty gravel and pebbles on low areas to the east and southeast. These gravel deposits, called the Tuscaloosa Formation, mark the beginning of late Cretaceous sedimentation in Tennessee (Stearns 1957: 1081–82). The gravel of the Tuscaloosa Formation is found as far east as the Western Highland Rim in Stewart and Dickson counties. More extensive deposits, found in the southeastern parts of West Tennessee in Hardin and Wayne counties, continue south and east into Northern Mississippi and Alabama.

Some of the gravel from the Tuscaloosa Formation can be found as deposits in ancient sinkholes now covered by the formation. Sinkholes of this sort, known as paleokarst, resulted when limestone from the Silurian and Devonian ages was exposed to weathering during the Mesozoic era. Quarry operations in the Parsons-Decaturville area have exposed in cross-section view a number of these Cretaceous age sinkholes filled with the gravel of the Tuscaloosa Formation.

Soon after the deposition of the Tuscaloosa Formation, a sea began encroaching upon the landmass. This sea, referred to as the Mississippi Embayment, is thought to have moved up from the south into West Tennessee. The axis of the embayment waters coincided with the present-day Mississippi River and extended as far north as Southern Illinois, Western Kentucky, and Southeastern Missouri.

The waters of the embayment were relatively shallow and teeming with life. The embayment received large quantities of sand and limey mud, sediments representing marine and transitional coastal environments not unlike the present-day Gulf Coast of Texas and Louisiana.

The sand is commonly cross-bedded; the limey mud is calcareous marl and chalk, locally very fossiliferous. The sand represents beach, sandbar, tidal channels, and possibly some deltaic deposits; the marl and chalk were deposited and formed in intertidal and subtidal marine waters (Stearns 1957: 1084–90; Russell 1965). Particularly noteworthy is the Coon Creek Formation, a greenish-gray, glauconitic, sandy marl, which is locally very fossiliferous in McNairy County.

Some Cretaceous clay beds and lenses probably were deposited in lagoons and swamps, as evidenced by the presence of numerous carbonized wood fragments and lignite deposits. Deposition of sand, silt, clay, and marl continued into the Cenozoic era, with the coastal environments oscillating back and forth as the embayment sea transgressed land and regressed.

Life during the late Cretaceous period was most fascinating and exciting, as the fossil record attests. The reptiles had by this time adapted to most environments, dominating air, sea, and land at the zenith of the dinosaur evolution on earth. The flowering plants, too, evolved during this time, significantly changing the vegetative cover of the land surface. In addition, two new mammal groups made their appearance during Cretaceous time: the marsupials, "pouched" mammals such as kangaroos, and the insectivores, resembling moles and shrews (Spencer 1962: 349).

The invertebrates also became very diverse during the Cretaceous period. Especially abundant were the unicellular foraminifera, mostly of microscopic size, and a group of extinct mollusks called ammonites, related to the chambered nautilus, were remarkably abundant. The Coon Creek Formation (so named for the type locality first described along the banks of Coon Creek in the Leapwood Community of McNairy County) bears a fossil fauna of mostly invertebrates that is renowned worldwide for their diversity and exceptional state of preservation (Fig. 24).

First reported by Bruce Wade in 1926, the Coon Creek fauna boasts over 350 different species of animals, including pelecypods (clams), gastropods (sea snails) cephalopod (mollusks similar to the chambered nautilus), foraminifera, crabs, and even vertebrate remains of mosasaurs (marine lizards that grew up to 40 feet in length) and turtles (Wade 1926: 191). Research on the fauna has revealed a complex

Fig. 24. One of the most unusual fossil occurrences is found at the McNairy County site of the Coon Creek Formation, where oyster shells such as this *Exogyra* can be found.

arrangement of paleoenvironments and animal/plant communities (Moore 1974: 142–67) that reflect intertidal and subtidal conditions.

Animal groups of particular interest because of the numbers of fossil specimens, quality of preservation, and ornate shell morphology include pelecypods (bivalve mollusks, such as modern-day clams), gastropods, and cephalopods. Unusually well-preserved specimens of crabs, including the pincers (Fig. 25), can also be found in the Coon Creek Formation (Wade 1926: 184–91; Moore 1974: 142–44).

The type locality of the Coon Creek Formation has been purchased by the Pink Palace Museum in Memphis. The museum conducts tours and special environmental camps at the rural site in McNairy County and provides special fossil digs for visitors (see Road Log, Mile 10.8, for directions to the museum in Memphis).

The disappearance of numerous life forms, including the dinosaurs, petrosaurs, marine reptiles (Ichthyosaurs and Plesiosaurs), ammonites, and belemnites (an extinct type of cephalopod that when fossilized has a distinctive cigar shape) marks the end of the Cretaceous period and the Mesozoic era. Ideas about the cause of the widespread extinction of

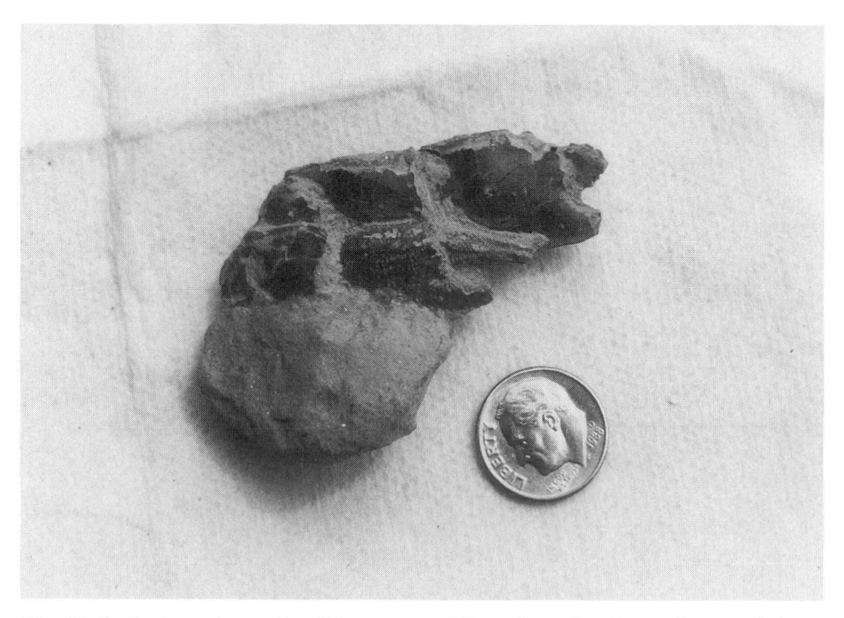

Fig. 25. Crab claws (over 60 million years old) are found in the sediment of the Cretaceous-age Coon Creek formation.

so large a group of animals have included not only starvation but also climate changes brought on by mountain-building processes. More recent theories involve the large-scale, cataclysmic impact of a comet, asteroid, or other meteorite into the earth. The results of such an impact, it is postulated, would raise enough dust particles into the upper atmosphere to sufficiently block out enough sun light to severely alter the earth's climate. Recent findings and continued research seem to support this idea.

Cenozoic Era

The Cenozoic era, also known as the Age of Mammals, is the geologic time period that extends from the present back to about 60 million years. It is during the Cenozoic era that major evolutionary changes occurred in the mammals. The flowering plants also diversified to major proportions, dominating the plant kingdom. Among the animals making their first appearances during the Cenozoic era were primates, rodents, horses, whales, and modern birds (Spencer 1962: 385).

In the Cenozoic era there are two main divisions: the Tertiary and the Quaternary periods. In Tennessee most of the Tertiary sediments are confined to the Gulf Coastal Plain province of West Tennessee. Quaternary sediments are found all across the state, including stream channels and floodplains as well as thick talus and block stream deposits along escarpment bases and high-elevation ravines of the Blue Ridge province.

Tertiary Period

In Tennessee the Tertiary period is represented by sand, silt, and clay of both marine and terrestrial origin. At the end of the Mesozoic era and the beginning of the Tertiary period, weathering of the Cretaceous strata in places resulted in an erosional surface. Sediment of the Tertiary sea covered most of the Gulf Coastal Plain of West Tennessee, forming an unconformity at the top of the Mesozoic strata.

The Tertiary period is divided into five epochs: Paleocene, Eocene, Oligocene, Miocene, and Pliocene, from oldest to youngest. Research indicates that the Paleocene and Eocene epochs record five advances and regressions of the Tertiary sea in Tennessee (Stearns 1957: 1090–99). As the sediments were deposited, subsidence of the Mississippi Embayment trough continued. The accumulation of this marine and deltaic sand and silt is found to be progressively thicker toward the embayment axis.

Clay deposits, resulting from deposition in swamps and lagoons along the seacoast, contain well-preserved leaf imprints and some lignite deposits. Lignite, sometimes called "brown coal," is formed from accumulation of organic matter in the same way as other coals are. Although it is still moist and has not hardened so much as bituminous coal, it is not so moist or soft as peat. Tertiary leaf fossils, particularly well preserved, can be found in the clay deposits around Puryear in Henry County.

A second unconformity, found in the upper Tertiary sediments, separates the Eocene and Pliocene epochs. Missing in this erosional period is approximately 24 million years of Tertiary sedimentation (including the Oligocene and Miocene epochs). The most recent epoch of Tertiary time, the Pliocene, is represented by gravel, sand, silt, and clay—sediments that appear to be remnants of alluvial deposits. Discontinuous across portions of West Tennessee, these deposits are thought to be sediments of ancestral streams of the present-day Tennessee, Ohio, Cumberland, and Mississippi rivers (Miller 1974: 49).

Erosion during the Tertiary period continued to excavate and enlarge the Central Basin. Escarpments of the Highland Rim and Cumberland Plateau were eroded to near their present locations during the close of Tertiary time. Prominent ridges and well-incised valleys, the distinctive features of the Valley and Ridge province, were well developed by the end of the Tertiary period. Erosion of the complex strata of the Blue Ridge during the Tertiary period has resulted in the well-worn sculpture of the Great Smoky Mountains, for example.

The fossil record of the Tertiary period in Tennessee is dominated by the numerous plant fossils found in West Tennessee. During Tertiary time the Mississippi Embayment, the sea across West Tennessee, was bordered by numerous lagoons and swamps where mud and fine silt accumulated (Stearns 1957: 1095–99). These plant fossils, many of which resemble modern-day plants, are preserved in the Eocene clay of West Tennessee.

Animal fossils, too, have been discovered in Tertiary age sediments. Turtle skeletal debris of Paleocene age was found near the Hardeman County community of Hornsby, and whale bones of Eocene age were found near the mouth of the Hatchie River at Fort Pillow in Lauderdale County (Miller 1974: 51).

Quaternary Period

The Quaternary period is composed of two epochs. The older, the Pleistocene, is widely known as the Ice Age. The younger, the Holocene, which represents the last 10,000 years of earth's past, is also called the Recent epoch.

During the Pleistocene epoch our planet experienced great changes in climate—from conditions of extreme cold, with snow and ice (called glacial periods), to warmer periods not unlike earth's climate today (called interglacial periods). In North America these extremes were also experienced as four major continental glacial ice sheets advanced across the landscape.

In Tennessee evidence of glacial activity has not been found. Research indicates that the ice sheets advanced as far south as the present-day Ohio River, along Kentucky's northern border. As the glaciers moved, the thick ice sheets carved the surface, pushing rock debris along in front of the ice (not unlike the way a roadgrader carves a flat roadbed, pushing up soil and rocks in front of the blade). As these glaciers started to melt, huge volumes of silt and ground-up rock debris (rock flour) were washed out in the melt water. These places where large volumes of rock debris pushed up by the ice were left, marking the farthest movement of the glacier, are called terminal moraines. These moraines are a prominent feature of the area near Kentucky's northern boundary.

Moraines are not found in Tennessee (because glaciers were not present in Tennessee during this time), but the Ice Age affected the Tennessee landscape in other ways. After the melting of the glaciers, winds picked up the silty rock flour and deposited it to the south. In Tennessee these deposits, called loess, are found in the western part of the state from the Mississippi River eastward to about midway across West Tennessee (see Road Log, Mile 82.2, and Side Trip 9 to Fort Pillow State Historic Area). The loess deposits are thickest along the Mississippi River, where the prominent bluffs are capped by 50 to 60 feet of this material. Eastward, the loess deposits thin to a few inches.

In East Tennessee the extreme cold during the glacial advances produced tundra conditions and even established a tree line along the Blue Ridge: virtually no trees appear above about 3,500–4,000 feet in elevation (Delcourt and Delcourt 1979, 1981, 1985). Without vegetation the higher slopes of the Blue Ridge were subjected not only to snow packs and snow fields but also to extreme freeze-thaw conditions. Natural rock bluffs were greatly affected by these conditions, developing thick talus, block streams and block fields in the lower elevations (Clark, Ryan, and Drumm 1987: 18–19). Today, if you visit the Great Smoky Mountains National Park in East Tennessee, you can see these boulders (some as large as cars and even small houses) in the stream channels. Originally these boulders were quite angular, but as they were transported away from their source they became subangular to subrounded and eventually some became rounded in shape.

Such deposits of soil and boulders (referred to collectively as colluvium) have posed great problems for highway construction in Tennessee, especially along I-40 in Roane County and I-75 in Campbell County. These colluvial deposits, which accumulated along the Cumberland Escarpment, are quite unstable. Some are masked by a thin veneer of soil and vegetation. As highway embankments were constructed over these areas, landslides developed, resulting in millions of dollars in damages (Royster 1973: 255–57).

Another feature of the Tennessee landscape that developed primarily in the Quaternary period is caves. Most of the present-day solution caverns and karst landforms found across Tennessee began developing within the last million years. Karst development was a form of weathering and erosion of the limestone of the Central Basin and Valley and Ridge before the Quaternary period, but most caves that developed in Tertiary time have also been eroded and weathered away.

When carbon dioxide from the atmosphere is dissolved in water, it produces carbonic acid; the acid, in groundwater, then dissolves limestone (producing calcium bicarbonate) as it percolates down through the soil and into cracks in the limestone. In some places this process has enlarged the cracks and fractures in the carbonate rocks (limestone and dolostone) to channels and openings big enough for humans to enter. Some of these solution cavities have subsequently enlarged into interconnecting cavities and developed into a labyrinth of passages (see Side Trip 6 to Cedars of Lebanon State Park). Today we call these large cave systems caverns. Exploration and research by the Tennessee Cave Survey, members of the National Speleological Society, and university specialists have resulted in mapped and cataloged caves numbering over 2,000 in the state (Barr 1961; Matthews 1971). A number of the larger caves, naturally decorated with cave formations (speleothems), have been commercially developed for tourism: Bristol Caverns (Sullivan County), Tuckaleechee Caverns (Blount

County), Indian Cave (Grainger County), Ruby Falls Cave (Hamilton County) and Cumberland Caverns (Warren County), for example.

Life during the Quaternary period seemed to be dominated by giant species of mammals. Large, elephant-like creatures called mastodons and woolly mammoths lived across Tennessee during this time. Other large animals included giant ground sloths, camels, saber-toothed cats, giant panthers, and jaguars (Corgan 1976: 23–35). The bones and teeth remains of these animals have been found not only in numerous caves (Big Bone Cave in Van Buren County and the Lost Sea in Monroe County, for example) but also in floors of sinkholes and depressions and in the floodplain deposits of old, sometimes terraced, beds of streams and rivers. The giant mammals now are gone. Perhaps the rapid change in climate approximately 10,000 years ago was too much for them to overcome. Perhaps, too, they could not adapt to the appearance of a new kind of predator on the continent: Homo sapiens.

During the most recent 10,000 years, Tennessee has experienced climate that ranged from temperate to humid subtropical; modern populations of both plants and animals have flourished. Today, humans seem to be the dominant species on earth. Certainly, in the last 200 years, humans have developed—for good or for ill—the technological ability to alter their environment in ways that no earlier species could.

Nonetheless, the earth's processes of vulcanism and mountain building, continental drift, glaciation, sedimentation, weathering, erosion, and evolution of life continue to operate as they have for hundreds of millions of years. Whether humans will be around long enough to see the results depends on their willingness to accept a role as a part of the natural process rather than as its antagonist.

Tennessee's Topography

When you look out over your yard and see that the land surface slopes down or up, or remains flat, you are looking at topography. If you look at the horizon and notice the silhouette of a mountain, the outline of a distant ridge, or a stream meandering across a valley, you are observing elements of topography. As you drive along a highway that sweeps down a long hill and then snakes back up a steep, narrow ridge, you are experiencing the effects of topography.

The configuration of Tennessee's land surface, including its relief and the position of its artificial and natural features, is known as Tennessee's topography. The topographic landforms vary—from the mountains, ridges, and valleys in the east to the flat, swampy lands in the west. High plateaus and rolling hills connect the east with the west to develop one of the most varied topographic areas of the Eastern United States.

Because Tennessee, in the Southeastern United States, occupies a long, narrow area, it is bordered by eight other states. Clockwise, these states are North Carolina, Georgia, Alabama, Mississippi, Arkansas, Missouri, Kentucky, and Virginia.

The approximate global location of Tennessee by latitude and longitude is latitude 36° 40' north to 35° 00' north and longitude 81° 40' west to 90° 00' west. With an area of approximately 42,250 square miles, Tennessee ranks thirty-fourth in size among the states (Klepser 1967: 73).

Tennessee's northern border with Kentucky and Virginia is 432 miles, and its southern boundary with Georgia, Alabama, and Mississippi is 336 miles. The greatest width, north to south, is approximately 115 miles, with an average width of about 109 miles. The distance in a straight line from Tennessee's southwest corner, near Memphis, to its northeastern corner, near Trade, is 483 miles.

The lowest point in Tennessee, 180 feet in elevation, is at the southwest corner of the state along the Mississippi River at Memphis. The highest point, 6,643 feet, is in the Great Smoky Mountains at Clingmans Dome.

Flowing down, around, and across the mountains, ridges, valleys, and swamps are numerous rivers that continue to shape the landscape and erode the rock formations. Among the largest are the Mississippi, Tennessee, Cumberland, Duck, Buffalo, Holston, French Broad, and Nolichucky. Other rivers, perhaps smaller but just as important in eroding and shaping the Tennessee landscape, include the Ocoee, Hiwassee, Little Tennessee, Doe, Obed, South Fork of the Cumberland, Sequatchie, Hatchie, Obion, and Forked Deer.

Of the many large lakes in Tennessee, all are artificial except one: Reelfoot Lake. With its origins in the catastrophic earthquakes of 1811 and 1812, Reelfoot Lake, in northwest Tennessee, serves to remind us of the power of our restless earth (Fuller 1912: 7).

From the shape of the land surface across North America, scientists have been able to distinguish six major physiographic divisions. The six major divisions have been subdivided into twenty-five provinces, and these further subdivided into sections.

The divisions, provinces, and sections are differentiated according to topography (how the land surface is shaped). The topography is influenced by not only the erosive character of the environment but also the character and structure of the underlying rock formations.

As you travel across Tennessee from east to west, you will encounter three major physiographic divisions: the Appalachian Highlands in the eastern part of the state, the Interior Lowlands in the middle, and the Atlantic and Gulf Coastal Plain in the west (Klepser 1967: 73–77; R. L. Wilson 1981: 5). The three grand divisions of the state—East, Middle, and West Tennessee—are derived from this physiographic differentiation.

The three major divisions are subdivided into five provinces and four additional sections, here collectively referred to as physiographic provinces. From east to west these nine physiographic provinces are the Blue Ridge, the Valley and Ridge, the Cumberland Plateau, the Eastern Highland Rim, the Central Basin, the Western Highland Rim, the Western Valley of the Tennessee River, the Gulf Coastal Plain, and the Mississippi River Alluvial Floodplain (Klepser 1967: 73–77; R. L. Wilson 1981: 5; see Map 5).

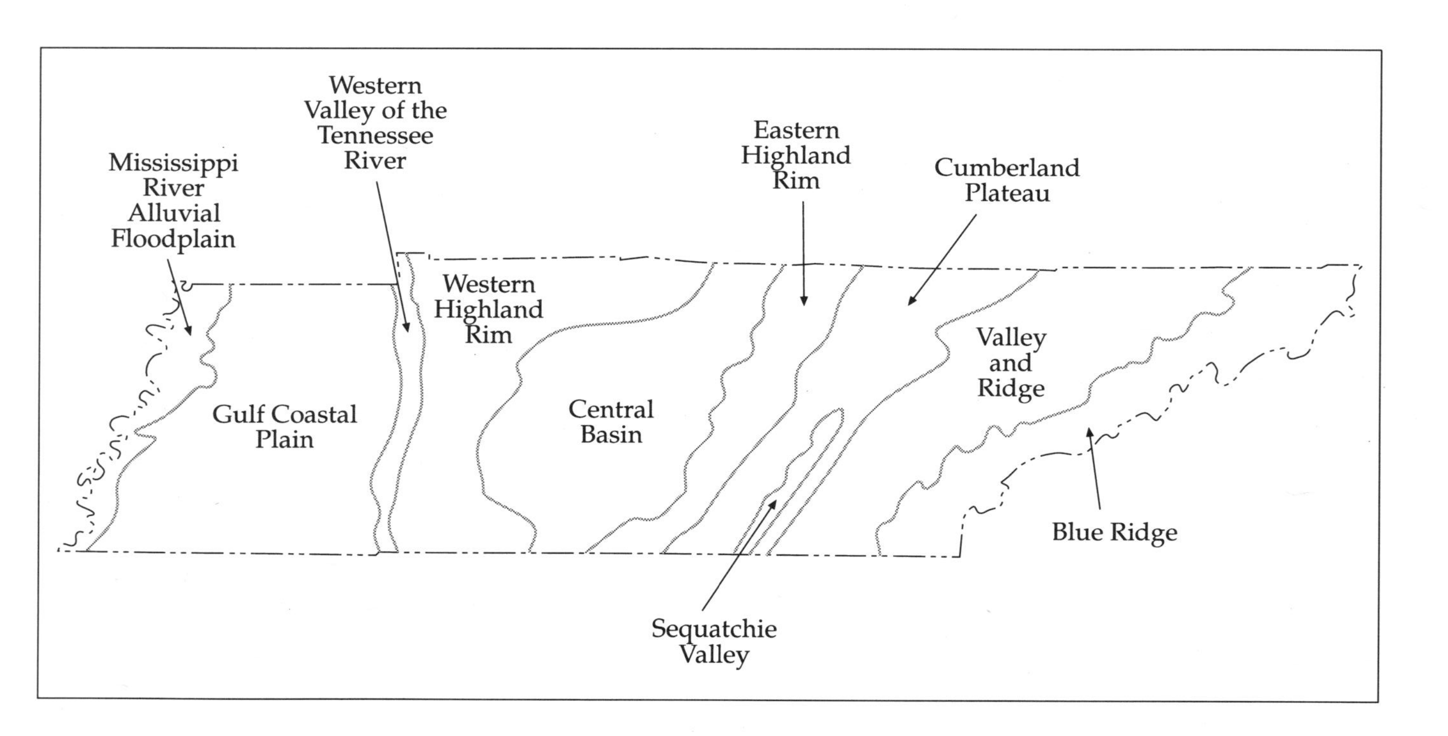

Map 5. Generalized physiographic map of Tennessee. After R. A. Miller 1974, 2.

Appalachian Highlands

The Appalachian Highlands division consists of landforms stretching from the Gulf Coastal Plain in the south to New England. Three of the seven provinces found in the Appalachian Highlands division are in East Tennessee: the Unaka Mountains (commonly called the Blue Ridge), the Valley and Ridge, and the Appalachian Plateau.

The general overall topographic expression of the land surface and the structural character of the bedrock in the Appalachian Highlands division run in an elongate northeast-southwest direction. Most major transportation routes follow this trend, including some portions of I-40 in East Tennessee.

Blue Ridge

The Blue Ridge province consists of the high mountains of easternmost Tennessee (Fig. 26). Sometimes referred to as the Unaka–Blue Ridge, this province contains several large mountains, including Chilhowee, English, Greene, Holston, Iron, Starr, and Roan. The Great Smoky Mountains are also included in the Blue Ridge (the name is correctly applied only to those mountains located within the Great Smoky Mountain National Park).

Fig. 26. The Blue Ridge Province is characterized by the high mountains of extreme East Tennessee, including the Great Smoky Mountains National Park.

The Unaka Range on the east and the Chilhowee Range on the west define the borders of the Blue Ridge in Tennessee. Elevations vary from 3,000 feet in the Chilhowee Range to over 6,600 feet in the Great Smokies, where there are twelve peaks that exceed 6,000 feet in elevation. The three highest peaks are Clingmans Dome (6,643 feet), Mt. Guyot (6,622 feet), and Mt. LeConte (6,593 feet). The present topography of the Blue Ridge is the result of millions of years of erosion. The ancestral mountains were much larger—in volume, elevation, and lateral extent.

Coves. Among the most interesting features of this province in Tennessee are the deep valleys and coves of the Great Smoky Mountains. These mountains are composed of metamorphosed sedimentary rocks of Precambrian age that have been thrust over younger Paleozoic sedimentary rocks (including limestone and shale) by a low-angle fault known as the Great Smoky fault. The coves were formed as a result of erosion of the older "thrusted" rocks, exposing the underlying ones, which are younger, softer, and more soluble.

Areas where the Paleozoic limestone and shale have been exposed have resulted in the formation of mountain coves such as Cades Cove, Tuckaleechee Cove, and Wears Cove. These coves, surrounded by older and topographically higher rock strata, form geologic "windows" providing a view down into the younger (limestone and shale) strata of the cove (see Side Trip 2 to Cades Cove).

Continued erosion of the surrounding mountain slopes has washed layers of soil and rock debris into the coves. This erosion, combined with the weathering of the limestone that forms the floor of the coves, has produced deep, fertile soil. It is not surprising that early settlers moving into this region often chose to settle in these coves, clearing the valley floors and farming the fertile soil.

Karst. Another feature of the coves that results from the weathering of the limestone is the development of karst, a landscape characterized by depressions, sinkholes, and caverns (Fig. 27). Karst occurs in areas underlain by limestone and other carbonate rocks. (Both East Tennessee and Middle Tennessee contain large areas of karst.)

Areas typified by karst usually, though not always, exhibit other features, such as numerous outcroppings of the bedrock (extensive exposures

Fig. 27. Karst areas are usually characterized by numerous exposures (outcrops) of limestone strata.

are known as lapies), sinking or disappearing surface streams (ponors), and large areas of coalescing sinkholes and sinking streams (uvalas).

The formation of karst begins as the breakdown of carbonate rocks (limestone, dolostone) by chemical weathering processes (Fig. 28). Rainwater that filters down through the soil and into the limestone bedrock is chemically changed to a dilute form of carbonic acid. This acid then attacks the limestone bedrock, dissolving the rock surface as it seeps down along cracks in the strata. As the solution activity continues, the cracks are enlarged until a sizable opening is developed. Solution-enlarged cracks continue developing and often intersect each other, as do the cracks in the bedrock.

Over long periods of time, the interconnecting solution cavities coalesce into larger and larger cavities, eventually becoming an underground maze of cavities called a cavern. Often, weathering at the surface intersects the subsurface weathering, forming sinkholes and depressions (Figs. 29 and 30). Sometimes the sinkholes have an open bottom called a swallet. These openings are "doors" to caves and caverns.

Fig. 28. Parameters that govern the development of karst-related problems may include: (1) Precipitation rates, (2) temperature, (3) vegetation, (4) geologic structure, (5) chemical content of strata, (6) joint/fracture development, (7) water table location, (8) soil type and thickness, (9) man's impact, (10) surface run-off, and (11) existing cavities.

Fig. 29. Sinkholes are a surface reflection of subsurface solution activity in karst areas; note man standing between two sinkholes.

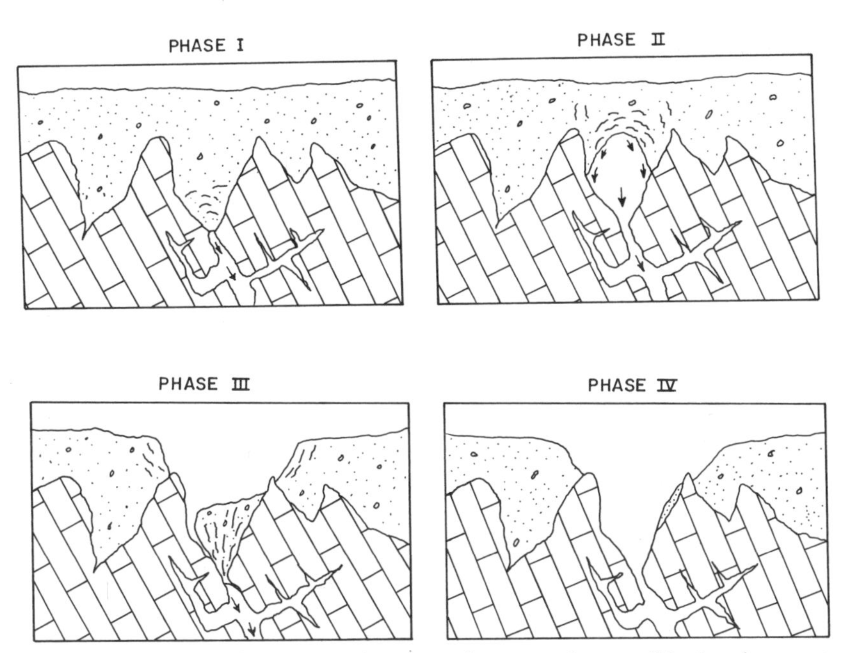

Fig. 30. This series of schematic diagrams illustrates the possible development of collapse sinkhole features in karst terrain. Phase I, soil with cavernous rock. Phase II, Soil with cavernous rock cavity. Phase III, Soil with cavernous rock, depression, and Phase IV, Soil with cavernous rock, sinkhole.

The environments of caves are intriguing. Ferns and mosses frequently live around the damp, cool openings of caves. Inside, animals such as mites, flies, crickets, spiders, and salamanders make their home, often feeding on organic debris washed into the cave ecosystem by a surface stream or collapsing sinkhole. Crayfish and bats complement the other forms of life in the complex cave environment.

The interior surfaces of caves are often decorated with formations (called speleothems) formed from the deposition of calcium carbonate by calcium-enriched groundwater. Some of the more common types of speleothems are stalactites, which hang down from the ceiling, and stalagmites, which grow up from the floor. Columns are formed where stalactites and stalagmites grow together; flowstone results when sheets of groundwater flow over a wall or floor of a cave, leaving the calcium carbonate deposit there.

Some caves have been developed for tourists, and others are "wild" (that is, not "improved," and thus requiring you to crawl and climb, get muddy, and provide your own source of light). Noteworthy among the various caverns in the coves of the Great Smoky Mountains are the commercial Tuckaleechee Caverns in Townsend and the "wild" Bull Cave in the Great Smoky Mountains National Park. (*Note:* You must obtain a permit from the park superintendent's office before entering any cave in the Great Smoky Mountains National Park).

Waterfalls. The lush, heavily forested mountains of the Blue Ridge have steep, rugged topography. Among the most beautiful features of the Blue Ridge are the rushing mountain streams and cascading waterfalls, such as Rainbow Falls, Ramsay Cascades, and Abrams Falls in the Great Smoky Mountains National Park (Table 9).

Waterfalls and rapids occur where there is a sudden drop in the stream channel. Generally, a resistant rock layer (sandstone, granite, or even limestone) forms the caprock over which a stream flows. Beneath this caprock is usually some layer or zone of softer or weaker rock, which will erode faster than the caprock.

Joints in the bedrock usually provide access to the weaker strata below, allowing the stream's water to penetrate and erode the weaker strata. The weaker strata, however, are not necessarily softer rocks; they may be more erodible because they were intensely fractured from past earth dynamics.

In time, the weaker strata are eroded away, leaving the harder caprock to form a ledge in the stream channel. During this process the

Table 9. Some Notable Tennessee Waterfalls

Waterfall	Province	County	Rock type	Rock age
Rainbow Falls	Blue Ridge	Sevier	Metasandstone	Precambrian
Abrams Falls	Blue Ridge	Blount	Metasandstone	Precambrian
Ramsay Cascades	Blue Ridge	Sevier	Metasandstone	Precambrian
Fall Creek Falls	Cumberland Plateau	Van Buren	Sandstone, shale	Pennsylvanian
Cane Creek Falls	Cumberland Plateau	Van Buren	Sandstone, shale	Pennsylvanian
Piney Creek Falls	Cumberland Plateau	Van Buren	Sandstone, shale	Pennsylvanian
Ozone Falls	Cumberland Plateau	Cumberland	Sandstone, shale	Pennsylvanian
Burgess Falls	Highland Rim	White	Limestone	Mississippian

overhanging ledge eventually breaks away, possibly forming a rubble-strewn channel that produces a cascade. Finally, the rubble erodes away, exposing the caprock and underlying weaker rock, forming a free-fall plunge in the stream course: a waterfall.

Natural rock arches. A most unusual feature of the Tennessee landscape is the occurrence of natural arches and bridges (described in greater detail in the section on the Cumberland Plateau, below). One of these rock structures usually looks like an arch of sculptured stone—similar to a span of a bridge. Rock arches and bridges, generally found to be products of weathering and erosion, occur in unusual and often delightful settings.

Unlike the rock arches of the landscape in the western United States, the natural bridges and arches in Tennessee are usually hidden under the cloak of our forest environment. Often ferns, mosses, and wildflowers occur in those locations. Fortunately for their preservation, these rock arches are often found in rugged or remote areas not suitable for economic or commercial development.

The presence of 36 natural bridges and arches in Tennessee has been documented and described (Corgan and Parks 1979). Two such rock structures in the Great Smoky Mountain National Park, however, were not mentioned in this context: Arch Rock and the Devil's Eye, both in the Alum Cave section on Mount LeConte. Arch Rock, under which the trail to Alum Cave Bluffs passes, is a 20-foot-high rock arch developed in a Precambrian slate formation. The Devil's Eye is a hole 5 feet in diameter through a very narrow outcropping of Precambrian slate.

Valley and Ridge

Located between the Blue Ridge on the east and the Cumberland Plateau on the west is an area commonly referred to as the Valley of East Tennessee. The Valley and Ridge province consists of numerous parallel ridges and valleys that trend in a northeast to southwest direction.

In lateral extent the Valley and Ridge province stretches from the Gulf Coastal Plain in Alabama northeastward to the Hudson Valley. In East Tennessee the Valley and Ridge is approximately 200 miles long and 45 miles wide.

Ridges in the Valley of East Tennessee tend to be long and narrow, with crests between 1,200 and 2,500 feet in elevation (Fig. 31). In many cases, a particular ridge maintains its elevation without much change along the crest. Some of the higher ridges—often referred to as mountains—include Clinch Mountain (2,485 feet) in Grainger and Hawkins counties; Bays Mountain (3,097 feet) in Greene, Hawkins, and Sullivan counties; and Whiteoak Mountain (1,495 feet) in Bradley and Hamilton counties.

Erosion processes have worked long and hard on the Valley and Ridge area. Because of differences in the hardness of the rock layers, some rocks have been completely weathered away while other strata

Fig. 31. The Valley and Ridge Province of East Tennessee consists of numerous parallel ridges and valleys. Some of the ridges are very high; for example, Clinch Mountain (foreground) and House Mountain (background) just northeast of Knoxville.

remain intact. This process, called differential erosion, has resulted in the formation of alternate ridges and valleys. The softer, more easily eroded rocks, such as shale and limestone, often form the valleys and lower slopes of the ridges (e.g., Poor Valley and Richland Valley in Grainger County), the more resistant sandstone and siltstone the ridge crests (e.g., Clinch Mountain).

Present-day streams, following the contours of the ridges and valleys, produce a drainage pattern in which the master streams in the valleys are at right angles to the tributary streams flowing down the ridge slopes. In this pattern, called trellised drainage, the streams flow down the ridges at right angles to the elongated ridge crests and then follow the valley floors to connect to larger, older streams that are parallel to the valley and the adjacent ridge. These older streams, emplaced before the present land surface was established, cut across the grain of the ridges in places, possibly following fracture patterns in the rock strata. In general, however, the drainage of the Valley and Ridge flows from areas in the northeast (Kingsport and Johnson City) to areas in the southwest (Chattanooga).

Karst. Because there are many locations with limestone strata in the Valley and Ridge, karst is a noticeable characteristic of the landscape here (Fig. 32). Numerous sinkholes, depressions, and cave systems dot the region (Figs. 33 and 34).

Of particular note is the Ten Mile Creek karst area in West Knox County. After flowing for more than a mile on the surface (passing beneath I-40), Ten Mile Creek sinks into a large cave system that is developed in Middle Ordovician limestone, finally reemerging in Fort Loudon Lake some 4,000 feet from the point at which it entered the cave system.

Commercial and residential development within the drainage basin of Ten Mile Creek has resulted in conditions that lead to flooding and water pollution. All the surface water that runs off from subdivisions and shopping centers, for example, empties into Ten Mile Creek. The creek then enters a constricted opening into the cave system. Flash flooding, which has always occurred in the area during heavy rains, has become increasingly worse in recent years. The runoff—with silt, grease, and tar from roads and parking lots and wastes from subdivisions and shopping centers—affects the groundwater conditions of the whole Ten Mile area. This situation illustrates the necessity for implementing zoning restrictions to control runoff.

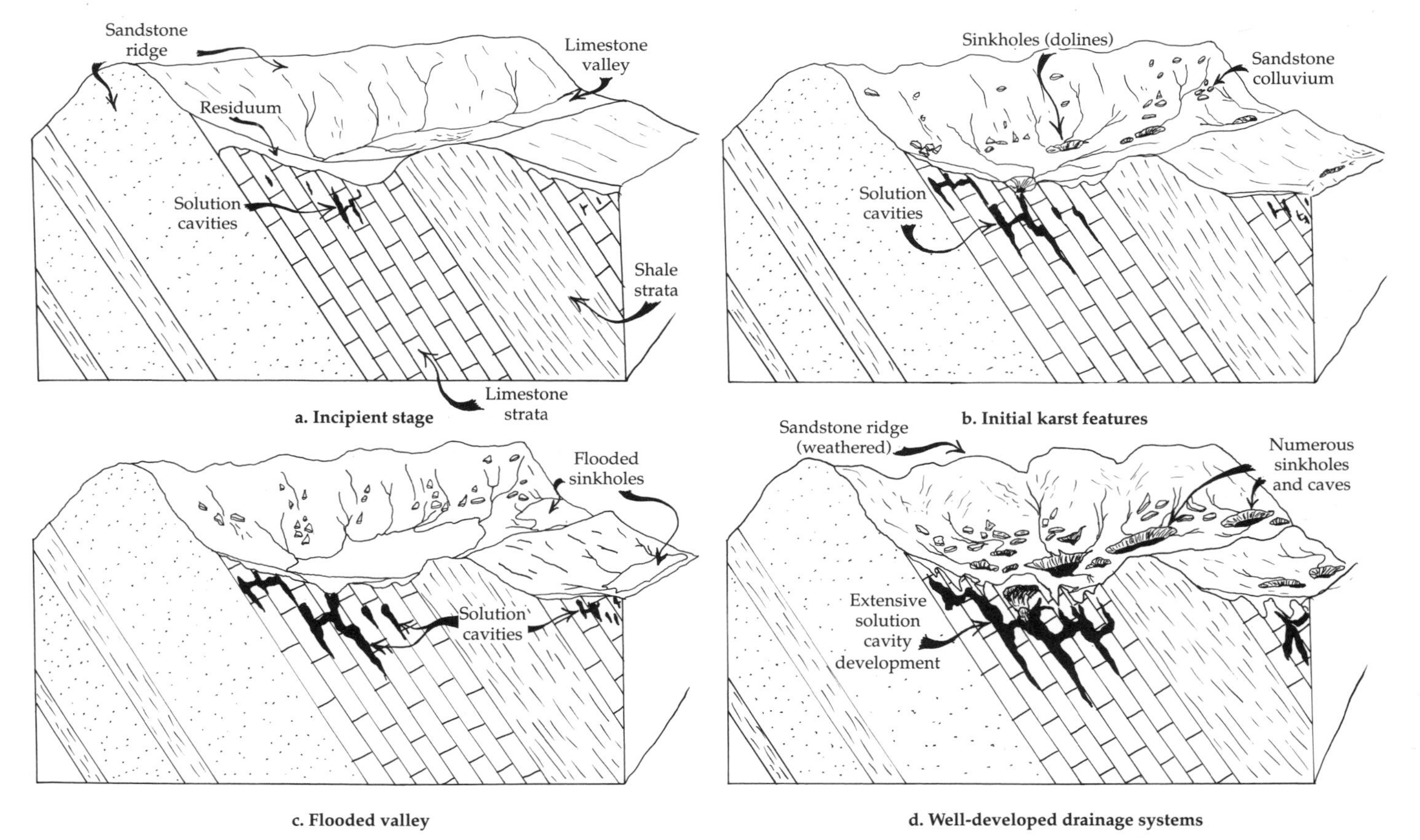

Fig. 32. This series of four schematic diagrams illustrates the development of karst in a hypothetical landscape. The four phases imply only gradual changes and not distinct events.

Fig. 33. The pollution of groundwater in karst areas occurs when trash is thrown in sinkholes as a means of disposal.

Fig. 34. Caves have intrigued people for thousands of years and are the object of enthusiasts called spelunkers.

Farther west, in Loudon County, I-40 crosses another section of karst (near Miles 367.0–365.2). Along this section of the interstate, several sinkholes have developed in recent years because of the activity of surface water that percolates down through the cavernous rock. This karst area, which is developed in strata of the Knox Group, extends along I-40 for about a mile and a half. (Some of the problems are mentioned in the section on maintenance in "The Making of a Highway.")

Cumberland Plateau

The Cumberland Plateau, a high tableland immediately west of the Valley and Ridge, forms the southern part of the Appalachian Plateau province (Fig. 35). The plateau area is bounded on the east by the Cumberland Escarpment, a long and steep demarcation between the Cumberland Plateau and the Valley and Ridge. Waldens Ridge is the name applied to the eastern plateau escarpment, which runs through Tennessee from Georgia to Virginia. The western escarpment, more irregular in shape, is also dramatic and scenic.

Fig. 35. The Cumberland Plateau is characterized by thick layers of very hard sandstone that commonly forms the caprock of the plateau, easily seen in this view of Cane Creek Canyon.

The Cumberland Plateau has an average elevation close to 2,000 feet. The southern two-thirds of the plateau is nearly level to rolling, but the northern third is broken by high mountains such as Crab Orchard Mountain (3,000 feet) in Cumberland and Morgan counties, Cross Mountain (3,550 feet) in Anderson and Campbell counties, and Round Knob (2,960 feet) in Scott County.

Sequatchie Valley. Two long, linear valleys are carved into the Cumberland Plateau: Elk Valley in the north and Sequatchie Valley in the south. These valleys are erosional features of a geological structure known as a faulted anticline, in which rocks have been folded upward in an arch and then broken and moved (in this case, northwest) along the length of the structure (R. L. Wilson 1981: 11). The erosion has been more pronounced along the broken, faulted rocks, incising a valley into the landscape. Although most of the rocks of the Cumberland Plateau are structurally flat-lying, some areas have been deformed as a result of the forces that brought North America in contact with Africa some 200 million years ago, the same forces that deformed the rocks of the Valley and Ridge and Blue Ridge provinces.

Sequatchie Valley, one of the most scenic valleys in the state (see Side Trip 4 to Fall Creek Falls State Park), is an impressive geologic feature (Fig. 36). Structurally, Sequatchie Valley is an eroded anticline with its long axis parallel to the long axis of the valley. The head of the valley is in Cumberland County while the body of the valley stretches southwestward through Bledsoe, Sequatchie, and Marion counties, extending southward into Alabama as Brown's Valley. The length of the valley is over 60 miles in Tennessee, and its width varies from 4 to 6 miles.

The steep walls of the valley, ranging from 1,000 to 1,500 feet above the valley floor, are an escarpment. Massive Pennsylvanian sandstone caps the plateau rim on either side of the valley, and Cambrian/Ordovician limestone is beneath the floor of the valley.

Elevations of the valley floor range from 600 feet near the southern end, where the Tennessee River cuts across the valley, to over 1,000 feet near the northern end of the valley beyond Pikeville. The rim of the valley extends to over 2,000 feet in elevation near the Alabama state line to over 3,000 feet near the head of the valley in Cumberland County.

The Sequatchie Valley's long history began near the end of the Paleozoic era, some 250 million years ago. A long, upward bending of

Fig. 36. Scenic Sequatchie Valley is the eroded crest of a long narrow anticline over 60 miles in length.

rock strata, known as the Sequatchie Anticline, was formed by folding and thrust faulting at the close of Paleozoic time (Milici 1967: 179).

During much of the Mesozoic era, erosion began cutting down the anticlinal ridge of rock capped by the Pennsylvanian sandstone. Once the sandstone strata were breached, the more soluble limestone began dissolving, lending itself to solution-type karst development. It is thought that Sequatchie Valley developed as a series of coalescing sinkholes (uvalas) that were eroded out to form the present-day valley (Lane 1952: 291–95; Milici 1967: 183).

Even today these processes are at work, as evidenced by the karst activity at the northern end of Sequatchie Valley. It is there that the Crab Orchard and Grassy Cove uvalas are located, along with an unnamed uvala—now all a part of Sequatchie Valley. In Grassy Cove are numerous sinkholes, sinking surface streams, and cave systems (the most widely known is Grassy Cove Saltpeter Cave). Surrounded by a high ridge, the beautiful rural farming cove is, in effect, a large-scale sinkhole.

Continued erosion of the valley during the Cenozoic era deepened and broadened the valley, as well as extending the northern end. The Sequatchie River developed during this time, eroding a channel in the weathered rock material. During its erosive work the river deposited numerous cobbles, some of which are now preserved as terraced alluvial deposits that can be found up and down the valley (Milici 1967: 187–88).

Waterfalls. Another notable feature of the Cumberland Plateau landscape is the presence of numerous waterfalls that have formed over prominent cliffs of the flat, sandstone cap of the plateau. These waterfalls are usually located along the irregular western rim of the plateau, where erosion has cut deep canyons into the tableland. At 256 feet, Fall Creek Falls, in Van Buren and Bledsoe counties, is the highest waterfall in the Eastern United States. Another especially beautiful waterfall on the Cumberland Plateau is Ozone Falls in Cumberland County, which drops 110 feet into a round plunge pool.

Natural rock arches. A particularly unusual feature of the landscape of the Cumberland Plateau is the formation of numerous natural rock arches (Fig. 37). In the northwestern section of the plateau in Fentress and Pickett counties are a number of these dramatic structures, at least eight of which have been identified and named. In the Big South Fork National River and Recreation Area is the largest of these (see Side Trip 3 to Twin Arches). Also known as Double Arches, Twin Arches has spans of 135 feet and 93 feet (Corgan and Parks 1979: 75).

Although these spectacular geological oddities can be formed in many different ways, weathering and erosion of the land surface and bedrock are the major forces involved. During the geologic history of our planet, rock formations have undergone stresses and strains that usually result in the development of joints (cracks along which no appreciable movement has taken place) within the rock mass. Erosional processes then act upon the rock formations along the joints by dissolving, abrading, and wearing away the rock.

The five major ways in which such structures originate are cave collapse, headward erosion, gravity, widening of a joint, and erosion of the neck of an incised meander. (1) Cave collapse, common in karst areas, involves the surface collapse of rock into a cave passage or chamber, with an arch of bedrock left at the surface, which itself may contain another collapsed area. (2) Headward erosion involves the erosion by a stream in the head, or upstream direction, of a valley or ravine, where the erosion continues through a ridge that is capped by a resistant rock (e.g., sandstone). (3) When gravity is the primary cause,

Fig. 37. Weathering processes have sculpted the Tennessee landscape into very scenic and sometimes unusual forms such as this rock arch in the Big South Fork National River and Recreation Area on the northern part of the Cumberland Plateau.

the arch is formed rather quickly—and usually by catastrophic means, such as a rockfall in which a hole is left in the remaining rock face after a block of bedrock falls out of it. (4) The widening of a joint usually involves a process that enlarges a crack (joint) in the bedrock, often running water over a long period of time. (5) In the case of erosion of the rock over an incised meander, an arch develops where a stream has cut a deep canyon into hard bedrock and, in response to a crack or change in rock type, cuts through a curve of the stream in the canyon, leaving a short arch of rock between the two sides of the canyon (Corgan and Parks 1979: 3–16).

There are three additional minor classes of origin for natural bridges. (1) Gravity-controlled movement involves an activity such as wave action along streams or coastlines, or even frost action, that moves a block of rock out of a rock face. (2) Stratigraphically controlled differential weathering is the erosion of two different rock types (e.g., sandstone lying on top of shale), in which the lower layer of rock is eroded more rapidly, leaving the upper layer spanning the weathered-out lower layer. (3) Faulting, too, may move two different rock types into contact with each other; because of differential weathering a rock arch may result from the erosion of the weaker rock type, but this kind of origin is not very common (Corgan and Parks 1979: 16–17). In some cases natural bridges and arches were formed by a combination of processes (Corgan and Parks 1979: 17).

Interior Lowlands

The Interior Lowlands division covers all of Middle Tennessee and includes two major physiographic regions: the Highland Rim and the Central Basin. The Highland Rim encircles the Central Basin, forming a continuous rim around it (much like the rim of a large flat plate with a broad shallow basin). Thus, in crossing Middle Tennessee from east to west, you encounter first the Eastern Highland Rim, then the Central Basin, and finally the Western Highland Rim. The Interior Lowlands division is underlain by sedimentary strata that are gently folded across the central portion of Tennessee—mostly limestone, dolostone, and shale that were originally flat-lying.

Eastern Highland Rim

The eastern section of the Highland Rim, adjacent to the western border of the Cumberland Escarpment, has a flat to slightly rolling terrain. The Eastern Rim is about 15 to 20 miles wide and averages slightly more than 1,000 feet in elevation. The highest point of the Eastern Rim is Short Mountain (2,074 feet), an erosional remnant of a once broader Cumberland Plateau (R. L. Wilson 1981: 15). Short Mountain, in Cannon County about 20 miles west of the Cumberland Escarpment, is also the highest topographic feature of the entire Highland Rim. Descending from the Eastern Highland Rim, you enter the Central Basin, an area encircled by the Highland Rim; this escarpment is the western boundary of the Eastern Highland Rim.

Waterfalls. The Eastern Highland Rim is dissected by deep erosional valleys, some of which have cascades and waterfalls. One of the most widely known is Burgess Falls (Fig. 38), just southwest of Cookeville, which has a plunge of 130 feet (see Side Trip 5 to Burgess Falls State Natural Area).

The Eastern Highland Rim is flat because the limestone strata underlying the area are almost flat or dip toward the east. Extensive and prolonged weathering of this limestone has produced a thick mantle of cherty, residual clay soil, often a reddish to reddish-orange color. In some places this soil is over 200 feet thick.

Karst. A prominent topographic aspect of the limestone-laden Eastern Highland Rim is its karst features: the numerous depressions, sinkholes, and caves resulting from chemical erosion of the limestone (Figs. 39 and 40). A large area of karst occurs from Putnam and White counties into Warren County (Cherry 1959: 63; R. L. Wilson 1981: 14). One of the largest caves in Tennessee—and, indeed, in the entire Southeast—is Cumberland Caverns, in Warren County. Over 30 miles of passageways have been mapped in the cave system, which is developed in the Monteagle Limestone near the Cumberland Escarpment.

Along this escarpment there are not only numerous caves but also several well-developed karst valleys (small valleys or coves formed by the solution and sinkhole collapse of limestone). Noteworthy among these is Lost Creek Cove, just southeast of Sparta in White County (Crawford 1987: 21). Because the Mississippian age limestone strata

Fig. 38. Burgess Falls is formed where Falling Water River flows from the Highland Rim into the Central Basin over resistant strata.

Fig. 39. Caves developed in limestone strata commonly develop secondary deposits of calcite called dripstone (stalagmites, stalactites), which decorate the cave interior.

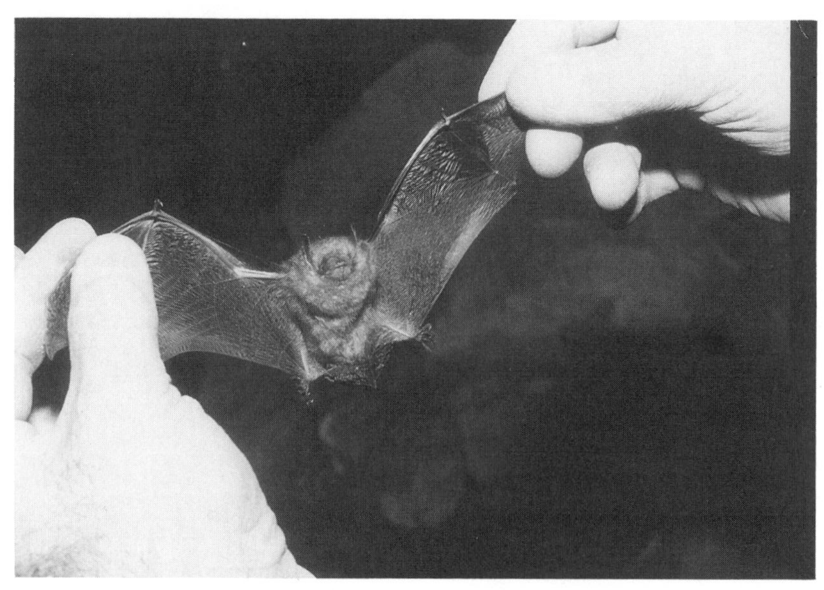

Fig. 40. Bats are flying mammals commonly found in the caves of Tennessee. They provide a natural method of insect control.

underlying this cove area have undergone extensive solution activity, the results include several large cave systems, many sinkholes, and a number of ponors (places where surface streams sink into the subsurface by means of sinkholes). Caves in the area—including Lost Creek Cave, Virgin Falls Cave, and Mill Hole Cave—have been mapped and studied by members of the National Speleological Society (Crawford 1987: 22).

In addition to geologic mapping and mapping of the cave systems themselves, scientific research into the development of Lost Creek Cove and its related caves has utilized dye tracings. This technique involves placing dye in surface streams before they sink into a cave and finding where the dyed water reemerges. Results of the various research efforts indicate that the caves carrying large amounts of water (conduit caves) were formed through subterranean invasion by surface streams along the retreating plateau escarpment (Crawford 1987: 40–41).

With so much limestone available, it is not surprising to find that crushed limestone and agricultural products are the main economic resources of this region. There are many large limestone quarries, several of which can be seen along I-40 just east of Cookeville. Some petroleum and natural gas production has been lightly established along the northern portions of the Eastern Rim (Klepser 1967: 75).

Central Basin

As its name implies, this topographic region is a broad basin in the central portion of Middle Tennessee (Fig. 41). The Central Basin is encircled almost completely by the Highland Rim, which lies 600 to 700 feet higher. The Central Basin is a rough oval with its long axis oriented northeast-southwest. This oval basin, approximately 100 miles long and 50 miles wide, has an average elevation of 500 to 700 feet above sea level. Numerous knobs located near the Highland Rim escarpment rise 200 to 300 feet above the surrounding basin floor. As in the case of Short Mountain, these knobs are erosional remnants of a much broader Highland Rim (Cherry 1959: 69).

The Central Basin is part of a geological structure called the Nashville dome, a southern extension of the much larger Cincinnati Arch, which is located primarily in Kentucky and Ohio (Klepser 1967: 75). The development of the topographic basin has its origin in the structure of the rock layers. Deformation of the rock strata occurred near the end of Paleozoic time, warping rock layers in Middle Tennessee into a large, gently sloping anticlinal dome.

Fig. 41. The Central Basin and Highland Rim (east and west) are characterized by relatively flat-lying rock strata.

When rigid material is bent up or down, stress is induced most intensely in the crest of the bend. When you bend a plank, for example, it will usually break at the crest of the bend. Rocks behave in the same manner. As rock layers are folded up or down, stresses quickly accumulate along the crest of the warp. In response to the stresses, fractures and joints (cracks) develop in the rock layers. The weaker areas in the rock (with respect to weathering) are where most of the fractures are located. These areas break down more rapidly in response to the weathering agents and thus erode more easily.

In the case of the Nashville dome, the rocks along the crest of the dome were more intensely broken than the layers located along the flanks. As a result, they were weathered more deeply and eroded more quickly. As these rocks were removed, a basin developed. Over millions of years of weathering and erosion, the basin enlarged to its present size of approximately 4,850 square miles (Klepser 1967: 75). In an ironic twist of geologic history, the Nashville dome has ended as the Central Basin.

Most of the rocks found in the Central Basin are limestone of Ordovician age. In the interior of the basin, the limestone has been subdivided into two major groups: the older Stones River Group and the younger, more shaly Nashville Group (R. L. Wilson 1981: 14). Other rock units of predominantly limestone and shale lithologies are found mainly along the perimeter of the basin; the numerous rounded knobs along the edge of the basin, for example, are capped by the Devonian/Mississippian age Chattanooga Shale and the Mississippian age Fort Payne Formation.

Cedar glades and karst. Soil derived from the weathering of the limestone in the basin is generally sandy clay. In many areas this soil is very fertile, providing a basis for extensive agricultural production. In some places, however, the limestone weathers to bare or nearly bare rock surfaces commonly called "glades" or "barrens." Red cedar trees have adapted quite well to these patches of thin soil and bare, platy limestone areas to form cedar glades, such as are found in Wilson County (see Side Trip 6 to Cedars of Lebanon State Park). One-fourth of the area of adjacent Rutherford County is covered by these cedar glades (Cherry 1959: 70). The cedar glades are also unusual for their numerous karst features.

Phosphate deposits in Tennessee occur mostly in the Central Basin, with large deposits in Giles, Maury, and Williamson counties (R. L. Wilson 1981: 15). The phosphate occurs from weathering of the brownish phosphatic limestone found in the Bigby-Cannon Formation in south-central Tennessee (Wilson 1981: 15). Phosphates are used mostly in the production of fertilizers. Phosphate mining activity in Tennessee has decreased in recent years, although it had an active history during the first half of the twentieth century.

Western Highland Rim

Located adjacent to the western part of the Central Basin and bounded on the west by the western valley of the Tennessee River, the Western Highland Rim is a rectangular area extending from Kentucky down to Alabama. The Western Rim has rolling terrain (more rolling than the Eastern Rim), dissected by numerous streams and several large rivers.

Extensive erosion has made the boundary of the Western Rim with the Central Basin highly irregular in shape. Elevations of 1,000 feet are common in the southern section of the rim in Giles, Lawrence, and Wayne counties but fall to around 700 feet in the northern portions of the rim in Houston, Montgomery, and Stewart counties. The rolling topography is broken by some flat areas in the vicinity of Lawrenceburg and Hohenwald. In general, however, the western boundary of the rim is rugged terrain. Unsuitable for mechanized agriculture, much of the area has been left in timber.

Three major rivers dissect the Western Highland Rim, developing nearly mature river valleys within its boundaries. The Duck, Buffalo, and Cumberland rivers all cut across the limestone and cherty clay soil of this region. I-40 crosses the distinctly developed river valleys of the Duck and Buffalo rivers in Hickman and Humphreys counties, respectively.

The Western Highland Rim is underlain predominantly by limestone strata of Devonian and Mississippian age. Cherty soil of the Fort Payne Formation caps most of the uplands in the southern half of the rim. The northern portions of the Western Rim are littered with outcrops of the St. Louis and Warsaw Limestones (R. L. Wilson 1981: 16). Stream valleys cut by the Duck and Buffalo rivers have exposed older rocks of Devonian, Silurian, and Ordovician age; these are found primarily in the southern half of this region.

Cryptoexplosive structures. One of the most unusual features of the Western Highland Rim is the Wells Creek structure, a nearly circular basin along the south side of the Cumberland River in Stewart and Houston counties near Cumberland City (see the Side Trip 7 to the Wells Creek structure and Dunbar Cave State Natural Area). This feature, thought possibly to be the result of either a meteor impact or a volcanic steam explosion, is called a "cryptoexplosive" structure because of its similarity to volcanic structural forms (Wilson and Stearns 1968: 165). Two other such craters known in Tennessee are the Flynn Creek structure (along the border of the Eastern Highland Rim and Central Basin), near Gainsboro in Jackson County, and the Howell structure (in the Central Basin), near Fayetteville in Lincoln County (Price 1991: 25–26).

Western Valley of the Tennessee River

Between the boundary of the Western Highland Rim and the Gulf Coastal Plain lies an unusually small geographic region of the state called the Western Valley of the Tennessee River. In the western part of the state, the Tennessee River turns northward along the Alabama-Mississippi state line and flows north across Tennessee for approximately 110 miles before entering Kentucky.

The Western Valley trends north-south and varies up to 20 miles in width. The Tennessee River has developed a floodplain that ranges in width from about 1.5 miles in Benton and Houston counties (near the Kentucky line) to approximately 3.5 miles in the south in Hardin County (R. L. Wilson 1981: 17). The valley walls, dissected by numerous small tributary streams, are extremely eroded. In the river valley floor, these streams have deposited sediments, some as old as the Pleistocene epoch and others as recent as the latest rain.

Most likely, the geologic history of this river valley extends back in time to the end of the Mesozoic era when the Cretaceous seas were beginning to retreat westward and southward. As the soft sediments of the Gulf Coastal Plain became a landmass, the river may simply have followed the contact between the harder Paleozoic rocks and the softer Cretaceous strata. Perhaps the northward course was the result of stream piracy, in which the Tennessee River was captured by a part of the Ohio River drainage (R. Wilson 1981: 17), thus diverting the flow of the Tennessee in a more northerly direction and eventually connecting it to the Ohio.

There is no doubt that the last Ice Age, which began over two million years ago, had an impact on this valley. As the ice sheets in the north advanced southward, sea level dropped, causing the river to erode and deepen its channel in response to the lower basin level. During the warmer interglacial periods, when the continental ice sheets retreated northward, most of the water that was locked up in the ice mass was returned to the sea, raising sea level. Correspondingly, as sea level rose, the river changed from erosional to depositional in character, forming broad floodplains and accumulations of sediment in the valley floor (R. L. Wilson 1981: 17).

The dramatic character of this river valley is best experienced by traveling I-40 across the valley. The most rewarding views are obtained by traveling from the west side of the valley near Sugar Tree to the east side near Cuba Landing.

Atlantic and Gulf Coastal Plain

The area encompassed by the Atlantic and Gulf Coastal Plain division in Tennessee extends across two physiographic provinces: the Gulf Coastal Plain and the Mississippi River Alluvial Floodplain. These two regions cover approximately one-third of the state of Tennessee, extending from the Western Valley of the Tennessee River westward to the Mississippi River.

Once covered by Cretaceous age seas, this area is mostly sand, silt, and marl—rich in fossil evidence of past life. This area is also rich agriculturally, its fertile soil producing all of the cotton and most of the soybeans and wheat grown in Tennessee.

Gulf Coastal Plain

The Gulf Coastal Plain consists of a broad, gently westward-sloping landscape characterized by rolling to flat topography. This province, 80 to 100 miles wide, extends from the western valley of the Tennessee River to the bluffs along the Mississippi River.

Ground elevations vary from 600 feet near the edge of the western valley to about 350 feet along the Mississippi River bluffs on the west. This section, covering nearly 9,000 square miles (Klepser 1967: 76), is drained by several large streams, including the Hatchie, the Forked Deer (North, Middle, and South Forks), and the Obion rivers.

The flat to rolling topography is broken by the wide swaths of bottomland of the larger streams that dissect the plain (Fig. 42). The rolling topography is more pronounced in the eastern third of the plain where north-south trending ridges reflect the general attitude of the gently westward-dipping strata.

Cretaceous age silt, sand, and marl are exposed in the eastern third of the Gulf Coastal Plain while clay, silt, and sand of Tertiary and younger age are found elsewhere (Klepser 1967: 76; R. L. Wilson 1981: 18). Local deposits of Cretaceous clay, sand, and gravel, which are quarried for construction products, constitute the major "mineral" resource of the area.

Loess. One other significant characteristic of the Gulf Coastal Plain topography is the presence of a surface layer of angular silt known as loess. This wind-deposited material, which was blown into this region from the west and northwest, may have originated as silt in outwash plains of the melting glaciers to the north (Cherry 1959: 82).

The loess deposits are thickest along the Mississippi River bluffs (65 to 70 feet) and gradually thin eastward, covering the western half of the region. A peculiar characteristic of loess is that the silt material, angular in shape rather than rounded, can remain almost vertical in surface exposures, thus resisting extensive erosion. The Mississippi River bluffs are composed of thick accumulations of loess. These loess bluffs, some more than 100 feet high, collectively form the western

Fig. 42. The Coastal Plain Province is the home of numerous swamps and cypress trees found along many of the rivers that drain the area.

escarpment between the Gulf Coastal Plain and the Mississippi River Alluvial Floodplain. This bluff escarpment is most pronounced at Memphis, at the Fort Pillow State Historical Area, and along the eastern edge of Reelfoot Lake State Park.

Mississippi River Alluvial Floodplain

Locally referred to as the Mississippi River Valley, this is a topographically flat, low-lying area, 10 to 14 miles wide. In this section, located between the Mississippi River and the loess bluffs to the east, elevations vary from 250 to 300 feet above sea level (lowest just south of Memphis).

The Mississippi River (Fig. 43) characteristically has developed numerous meanders in its lateral movements back and forth within its valley. In this province the river exhibits other "old-age" characteristics—such as oxbow lakes, swamps, natural levees, and meander scars—indicative of a stage in a stream's development in which the processes of erosion are decreasing in vigor. These features can be seen all along the channel of the Mississippi but are particularly visible at Fort Pillow in Lauderdale County (see Side Trip 9 to Fort Pillow State Historic Area).

Fig. 43. The Mississippi River forms the western boundary of Tennessee, where loess deposits form bluffs overlooking the river. This view is looking upstream on the Mississippi River at Fulton, Tennessee.

One of the most dramatic and geologically significant places in this section of Tennessee is Reelfoot Lake. Located in the northwest corner in Lake County, Reelfoot Lake is the only major, naturally occurring lake in the state. The lake is the result of a geologic catastrophe—the violent earthquakes in December 1811 and January 1812, which occurred along the nearby New Madrid fault (see Side Trip 8 to Reelfoot Lake State Park). Because the intense vibrations from the earthquakes caused large areas of the area to settle, a large, shallow depression formed and quickly filled with stream runoff. Today this lake is enjoyed for its scenic beauty and its wealth of plants and wildlife.

Geologic Environmental Issues

A landslide, a sinkhole collapse beneath a road or building, a polluted water well, or toxic leachate emanating from a landfill site—you may have experienced one of these environmental problems either directly or indirectly (Fig. 44). Many environmental concerns facing communities across the United States involve conditions related to geology. In Tennessee particular environmental problems that are controlled or influenced by geologic conditions include landslides, sinkholes, groundwater pollution, solid waste disposal, mining reclamation, and engineering geology along highways.

Where you build a house, locate a highway, drill a well, or dispose of waste should be dependent upon and dictated by what geologic conditions are present. A house built in or over a sinkhole or in an area prone to sinkhole development may eventually collapse (Fig. 45). A highway in an area prone to landslides may be expected to experience an unstable future. A well drilled into rock strata transmitting groundwater from potentially hazardous waste sites should be expected to yield contaminated water. An understanding of the geologic conditions involved with such conditions, however, can help prevent the occurrence of potential problems.

Some of the general geologic conditions usually evaluated with regard to environmental geologic problems include topographic conditions, kind of soil, depth of soil, kinds and locations of surficial deposits (alluvium, colluvium), surface drainage, surficial expressions of instabilities, type of bedrock, attitude of bedrock, and physical and

Fig. 44. Highway-embankment landslides result where roadway fills are constructed on top of unstable material such as colluvium.

Fig. 45. Sinkholes occur along highways where cavernous limestone is encountered.

chemical characteristics of the bedrock. Site conditions and proposed use requirements may necessitate auger and/or core drilling and laboratory testing of soil and/or rock samples. Some of the common geologic environmental problems in Tennessee are detailed below.

Landslides

One of the most dramatic geologic-related problems is a landslide. The movement of earth, rock, and surface vegetation down the slope of a land surface can be a frightening sight. Houses crushed by moving earth, highways torn apart by sliding soil and rock, and mountainsides scarred by repetitive sliding have all occurred in Tennessee.

Landslides can occur naturally or be induced by human activity. Naturally occurring landslides usually result from a combination of heavy precipitation and weaknesses in rock and soil that have been brought about by weathering (Fig. 46). Human activity (such as housing developments, highways, railroads, and strip mines) frequently disrupts the natural stability of the land surface. Quite often human-induced landslides result in extensive economic damages, such as the destruction of roadways and houses.

Fig. 46. Landslides often block highways; some can be catastrophic, such as this one, which covered a tunnel entrance on I-40 just inside North Carolina.

Most landslides, whether natural or human-induced, occur in response to heavy precipitation over a short period of time. In 1973, for example, heavy spring rains set off numerous landslides along I-40 in Roane County along a part of the Cumberland Escarpment referred to as Rockwood Mountain (Royster 1977: 138).

Although there are many circumstances under which a landslide can occur, one condition that is universally required is a sloping land surface. Recent landslides in Tennessee (most of which have been human-induced) have been documented to occur in clay and sandy soil, colluvium, and masses of bedrock (Royster 1973; Aycock 1981; Trolinger 1975; Moore 1986). Two major types of landslides occur in Tennessee: translational (a category including wedge-failure landslides [Fig. 47] and block-glide planar slides) and rotational (slumps). Two other related types of movement, rockfalls and mudflows, have also been documented as taking place in the state.

When translational slides occur, the sliding mass progresses out—or downward and out—along a gently undulating or more or less plane surface. There is no rotational movement involved. One of the most

Fig. 47. Wedge-failure landslides closed I-40 in Cocke County, Tennessee, many times, some for as long as two weeks. TDOT photo by George Hornal.

common types of translational slides is the wedge failure (Fig. 48), which occurs as a result of the movement of masses of rock which slide along the line of intersection of two planes of discontinuities, usually fracture and bedding planes. Within seconds there is a thunderous roar of tons of rock, soil, and vegetation sliding onto the highway and leaving a wedge-shaped scar on the landscape above the road. The excavation of nearly vertical cutslopes for the roadway, in effect, undercut or "daylighted" these lines of intersection, permitting the rock wedges under certain conditions to move freely downslope.

Another type of translational landslide, called a block-glide planar failure, can be just as devastating as a wedge failure. Planar failures are commonly referred to as block glides because the failures usually result in blocks of rock sliding down a sloping surface. This slip surface is a plane, usually parallel to the surface of the ground. Planar-failure landslides occur when layers of gently dipping rock are cut into, exposing the dipping plane. As a result, the exposed layer of rock slides into the excavated area. This type of landslide can occur naturally, as when a river carves into a hillside and undercuts the dipping rock strata, or be induced by human activity, such as cutting a road into the side of a hill, which produces the same effect.

Several block-glide landslides have occurred along U.S. 25E on Clinch Mountain in Grainger County as a result of highway construction (see Side Trip 1 to Cumberland Gap National Historic Park). Repair work was time-consuming and costly (several million dollars).

The other major category of landslides, the rotational, usually occurs in masses of soil and moves along a surface of rupture that is curved upward (concave). Slumps, a common variety of rotational failure, involve movement about an axis that is parallel to the slope. Rotational failures may be either naturally occurring or human-induced. The former, a result of weathering and erosional processes, often appear as slumps of earth along soil bluffs and escarpments or sharp hills (see Side Trip 9 to Fort Pillow State Historic Area). The latter are common in homogeneous materials such as highway embankments and fills.

Of the two other related types of movement, among the most common are single-rockfalls. A single-boulder rockfall usually occurs as a result of the weathering of fractured rock occurring along precipitous roadway cutslopes. Highly joined rock strata combined with overblasting during construction makes rockfalls both common and dangerous. Many rockfalls in the mountains also occur when water

A. Rock slope as constructed

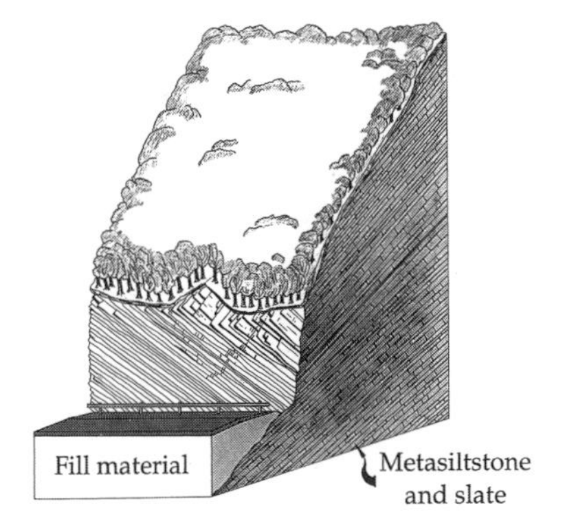

B. Incipient wedge failure

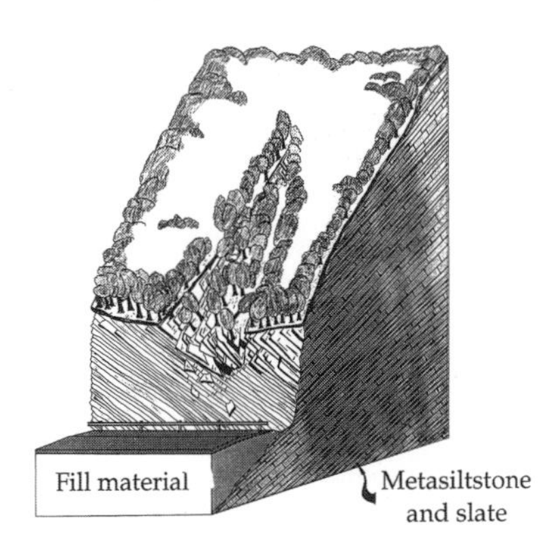

C. Rock wedge during failure

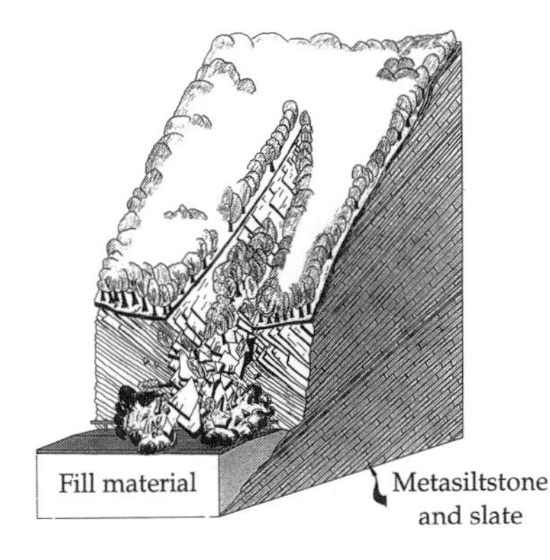

D. Rock slope after wedge failure

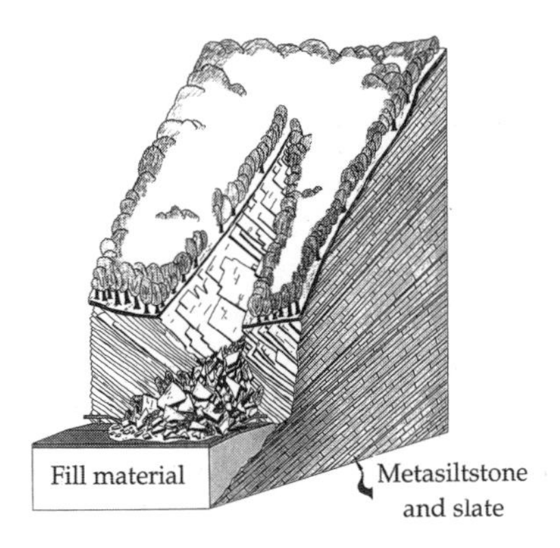

Fig. 48. A wedge-failure landslide (Moore 1986).

A. A natural-rock slope or an artificial roadway-cut slope, as along I-40 in Cocke County, as illustrated above, may have the potential for wedge-failure landslides.

B. The incipient development of a wedge failure may include cracks opening in the ground and trees leaning or tilting in response to ground movement. Episodes of rockfall may also be an indicator.

C. A wedge-failure landslide during maximum movement results in a very powerful force that destroys everything in the path of movement, usually in a matter of seconds.

D. The results of a wedge-failure landslide are a large wedge-shaped scar (exposing bedrock) on the landscape and a large mound of rubble and debris at the bottom of the slope.

that has seeped between strata freezes and forces rocks to fracture. While rockfalls involving large boulders are obvious hazards to motorists, many falls are of much smaller rocks. Indeed, the colluvium and talus slopes at the base of mountains are the results of multiple rockfalls.

Three particular sections of I-40 in Tennessee have been the subject of rockfall and landslide problems. All three are in East Tennessee: in Cocke County (Road Log, Miles 451.8–448.2), in Roane County (Miles 346.5–341.5), and in Cumberland County (Miles 333.4–331.4).

Another type of movement, the mudflow, also occurs in Tennessee. Mudflows, which involve high water content and move fluidly, range from small (10 to 20 feet high and wide) to very large (several hundred feet in height and a quarter of a mile or more in length). Most mudflows appear after long periods of heavy precipitation.

Karst

Karst is a term applied to an area of land characterized by surface depressions, sinkholes, caves, and underground drainage. Because the bedrock in karst areas is typically carbonate in composition (limestone and dolostone), it lends itself to chemical solution. As cracks in the bedrock are enlarged by solution activity, the cavities interconnect, collapsing into a series of open chambers commonly called caves.

Surface water percolating down through the soil and cracks in the rock ends up in the solution cavities, sometimes producing large underground streams. These underground streams are often tapped by well drillers as a source of potable water.

The surface sinkholes and depressions that receive precipitation runoff, which filters down through the soil and rock into the cavities and groundwater, thus serve to replenish the groundwater supply. Persons who dispose of waste materials into surface depressions and sinkholes pollute the groundwater supply with the very wastes that they were trying to dispose of: rainwater then runs into the trash-filled sinkholes, percolating down through the trash into solution cavities in the bedrock that carry the water to wells and springs.

Such pollution is unfortunately commonplace throughout karst areas in Middle and East Tennessee (as well as in other states like Florida, Kentucky, Indiana, and Alabama). In most instances, the people involved do not realize that they are polluting their own water source.

Another source of groundwater pollution in karst areas is non-point source pollution from runoff in urban areas. As the runoff picks up residues of tar, oil, gasoline and other toxic materials from parking lots and streets, surface streams carry these substances into sinkholes and ponors (places where surface streams empty into sinkholes, such as the Ten Mile Creek area in West Knox County). As a result, the groundwater becomes contaminated with the toxic substances.

Other problems experienced in karst areas include subsidence, collapse, and flooding. Although these problems can occur anywhere karst is present, they are usually not noticed until houses, commercial and industrial buildings, and roads begin to be affected (Fig. 49).

Sinkhole collapse and subsidence can be either naturally occurring or human induced. Naturally occurring sinkholes are those that develop without human influence, usually in remote rural areas. Human-induced collapses and subsidence are usually the result of humans having disrupted the naturally developed karst areas by building subdivisions, shopping centers, and highways on this cave-prone land. Grading of the surface soil alters the percolation of surface

Fig. 49. Karst areas can cause trouble for humans when structures are built in cave- and sinkhole-prone areas such as this sinkhole in South Knoxville.

water down through the soil layers. In most instances, grading operations bring surface water nearer to limestone bedrock (pinnacles), resulting in rapid erosion and infiltration into the subsurface. The net result is the occurrence of subsidence and collapse problems.

In actuality, the subsidence or eventual collapse results from the subsurface erosion of the soil where percolating water in the soil carries small particles of the soil downward into the underlying solution cavities. Over a period of time, these areas of subsurface soil erosion develop cavities within the soil mass. The soil cavities enlarge through continued erosion until the remaining soil above the cavity starts to sink and eventually collapses into the underlying cavity to form depressions and sinkholes. In a few rare instances, erosion of the bedrock is so extensive that the bedrock actually collapses into the cave below. Some of the problems that have resulted from humans doing their development over karst terrain include subsidences beneath house foundations, the collapse of yards because of leaking swimming pools and inappropriately located septic tanks, the collapse of highway surfaces and ditchlines, and numerous instances of flooding. However, because karst underlies so much of Middle and East Tennessee, highways sometimes cannot avoid it (Fig. 50).

Fig. 50. Highways are occasionally built through karst areas and experience sinkhole problems as a result.

Scientists have been studying the effects of humans on karst areas. The occurrence of sinkhole collapse features in Montgomery County, Tennessee, for example, has been analyzed and explained (Kemmerly 1980). Particular karst problems associated with Tennessee highways have also been documented, with descriptions given of numerous engineering corrections to these problems (Moore 1980, 1987, 1988). And there are many other studies on the development of karst areas and the engineering of related karst problems (Royster 1984; Newton 1976; Crawford 1987; Miller and Maher 1972).

Groundwater Pollution

Groundwater, the water in the soil and bedrock, usually comes from the percolation of surface water down through the soil and into the bedrock. People most often come in contact with groundwater by way of domestic water wells and springs. Springs are places at the surface where groundwater seeps from the subsurface bedrock into the surface environment.

Vast amounts of water are stored in the subsurface of the earth's upper crust in the form of groundwater. Most groundwater is replenished (recharged) by precipitation that infiltrates down into the subsurface. As the surface water flows through the soil and bedrock, it can pick up particles of clay and other substances, such as minerals (e.g., calcium carbonate and iron). Heavy mineralized water is often referred to as "hard" water.

The pollution of our groundwater supplies has occurred over the years, first through ignorance, but more recently because of laziness, irresponsibility, and, quite often, greed. In the earlier years of this century and even today, one of the most direct causes of groundwater pollution was and is the disposal of garbage waste into sinkholes. Whatever is placed into sinkholes can (and often does) pollute the water that flows through it and on into the subsurface.

Household garbage, dead farm animals, motor oil, and other materials placed in sinkholes provide point sources of pollution of the groundwater. Some of these wastes are often toxic and extremely hazardous. When buried in the soil at a landfill, these substances provide the same source of groundwater pollution, though at a slower rate. In the past undesirable waste was often disposed of by injection wells. This method, now outlawed, involved drilling a hole down into the bedrock and disposing of industrial and manufacturing waste into the hole. Out of sight, out of mind—no problems, it was thought. The result, however, was direct pollution of the groundwater supplies.

Groundwater pollution can occur from a number of other sources and geologic conditions. Use of unsuitable sites such as floodplains, swamps, and rocky ground for landfills to dispose of waste often results in pollution of not only subsurface water but also surface water supplies. Leakage of toxic or hazardous chemicals from point sources such as storage tanks or leaking pipelines can pollute the local water supplies.

Particular geologic factors can contribute to the pollution of groundwater supplies: for example, depth and composition of soil cover (in particular, thin sandy or silty soil), type of rock (including limestone, sandstone, or shale), porosity of rock and permeability of rock (ability to transmit fluids). Fluids can be transmitted through rock strata by not only granular porosity (between grains, such as in sandstone) but by cracks (joints, cleavage), bedding planes, and fault zones.

Complex groundwater conditions have been studied in a number of areas across the state of Tennessee. Detailed analysis of the geohydrology of the U.S. Department of Energy's Oak Ridge Reservation in East Tennessee relates the flow of groundwater to rock type and structural control (McMaster 1991). There has also been research on the relation of groundwater patterns to joints and other structures in the limestone strata found in the Upper Stones River Basin of Middle Tennessee (G. Moore, Burchett, and Bingham 1969). A study of the Skillman Basin in Lawrence County, Tennessee, documents the weathering characteristics and the geohydrology of Mississippian age limestone on the Western Highland Rim (Stearns and Wilson 1971).

One of the most interesting sites for studies of groundwater and pollution control is Mammoth Cave National Park, Kentucky (Quinlan 1983). The researchers have used extensive groundwater dye tracings of surface and cavern streams, including spring and ponor locations. Development of an extensive and accurate map of the groundwater movements in the Mammoth Cave area, including times required for water to flow between known points such as sinkholes or springs, has enabled park officials to activate appropriate mitigation measures at the necessary locations in the event of hazardous or toxic spills from tanker truck accidents.

In Tennessee, contamination of groundwater supplies is of particular concern. At the well field supplying the town of Alamo in Crockett County, for example, water sampling and testing of that particular supply area by the Tennessee Department of Health and Environment revealed the occurrence of toxic volatile organic compounds (Hutson 1989: 7). These compounds, including trichloroethene, dichloroethene, and carbon tetrachloride, were found in concentrations that exceed both state and federal standards for drinking water (Hutson 1989: 7). This problem, under study by the U.S. Geological Survey along with state and local personnel, points to the widespread problem of groundwater contamination even in rural areas of Tennessee.

Landfills

Another area of environmental importance and concern involves landfills—those unsightly excavations where our garbage is finally laid to rest. Because our society has a "throwaway" attitude—and a technology that supports it—tons of trash are generated every day in each community. The accepted means of disposing of this garbage is taking it to a landfill, where it is placed in an excavated trench and covered with soil.

In May 1969, the Tennessee Legislature passed the Tennessee Solid Waste Disposal Act (Tennessee Code, Sections 53-4301 through 53-4315), which requires closing all open dumps, prohibits unregulated open burning, and prohibits placing waste in state streams (Miller and Maher 1972: 2). In effect, this law requires the disposal of garbage in sanitary landfill operations meeting standards adopted by the State Department of Public Health.

Obviously, geologic factors play a very important role in the successful development and operation of a landfill. In landfill siting, surface topography, soil conditions, rock conditions, and flow of groundwater in and through the landfill site need to be considered (Miller and Maher 1972: 3–19). The topographic considerations include the slope and configuration of the surface, the stability of the land surface, how much elevation change (relief) is found at the site, and special topographic considerations such as the presence of karst, floodplains, and swamps. In addition, certain aspects of the soil (regolith) need to be determined: for example, the type of soil, its porosity and permeability, its shear strength, its moisture content, and the depth to bedrock. Some of the bedrock factors also need consideration: the type of rock (lithology), the attitude of strata (e.g., flat-lying or tilted), the degree of weathering, fracture and joint patterns, the groundwater table, the location of structural faults and folds, and the general porosity and permeability of the rock. Because the flow of groundwater through a landfill site can be influenced by these factors, all should be thoroughly understood and mapped out.

Defective landfills can be disastrous if toxic or hazardous leachates seep into the water table. A current study by the U.S. Geological Survey (U.S.G.S.) is investigating the nature and extent of groundwater contamination caused by the Shelby County Landfill in the Memphis area. An earlier investigation by the U.S.G.S. revealed that "downward leakage is occurring from the alluvial aquifer north and northeast of the landfill and that contaminants from the landfill are moving in the groundwater towards that area" (Parks 1990: 9). Leakage of contaminants can occur at almost any landfill site; in most instances, when leakage occurs, groundwater supplies are affected.

Strip-mine Reclamation

The type of mining in which the surface vegetation is removed and the soil and rock near the surface (called overburden) are stripped away in order to reach and extract a particular mineral or ore is called strip mining. Deep mining, in contrast, is activity that is underground, requiring tunnels and shafts to reach the deposits.

In Tennessee, strip mining is usually an effort required to extract coal deposits near the surface. Because much of the coal is in the mountains, the construction of mines along hillsides involves many of the same problems as building highways through that terrain.

Years of extracting coal by strip mining have in some places produced extreme environmental damage. Because for a long time control over strip mining activity was either nonexistent or not enforced, thousands of acres of mountain land in Tennessee were ripped apart and laid bare. Huge amounts of earth and rock (mine spoils) were shoved over the sides of mountains during the process of getting to and extracting coal. The scars of this past activity can be seen in the many strip-mine areas of the Cumberland Plateau.

Since the rock strata are mostly flat-lying in these areas, the coal seams are also horizontal. Accordingly, when strip mining occurs, there are horizontal strips of excavation around mountains and hills. In the past, these stripped areas were left open—bare as they were after the coal was removed. The consequences were erosion and siltation. Some of the mine spoils became saturated and broke apart, sliding down the sides of mountains. In addition, strata containing iron sulfide minerals deteriorated, resulting in the formation of acid runoff (acid mine drainage).

The landslides, erosion and siltation, and acid mine drainage had detrimental effects on the plant and animal life. The landslides further scarred the mountainsides, the siltation choked nearby streams, and the acid mine drainage lowered the water pH levels enough to kill the life in the streams.

Over the past twenty years, much time and effort has been devoted to the reclamation of strip-mined areas. This work includes landslide correction, restoration of the mined surfaces to their original contours, erosion control, reestablishment of vegetation, and remediation and control of acid mine drainage.

Increased attention to the stability of mine slopes and reclaimed areas in order to prevent landslides has generated particular interest in developing remedial designs, some of which are also useful to highway builders. Six of the most widely used methods for determining stability have been evaluated in order to document the analysis and design procedure better (Bowders and Lee 1990). Using these analysis concepts helps in correcting and preventing landslides.

Reestablishing plant and animal ecosystems not only improves the appearance of formerly mined land but is also one of the primary means of stabilizing the land. Considerable attention and research has concentrated on types of vegetation that are compatible with soil of strip-mine areas. Even the microbe contents of the mine soil are studied in an effort to produce the necessary nutrient and microbe interactions to support natural plant and animal successions (Zak and Visser 1990: 83). One study indicates that the slightly acidic sandy soil and sandstone rubble, commonly found in strip-mine sites, can be used to support grasslands or forests (Torbert and Burger 1990: 273–75).

The problems with acid drainage from strip-mine areas include the destruction of water habitat by acid water chemistry and the deposition of iron hydroxide (informally called "yellow boy"). Streams affected by this condition often become "sterile," supporting no plant or animal life. This condition can extend for miles downstream from the mine source, destroying complete ecosystems in the process.

Correcting and eventually controlling acid drainage has been the focus of some much-needed research (e.g., Brady and Hornberger 1989; Byerly 1981). The sources and geochemistry of the development of acid drainage, for example, have been examined through acid-base accounting methods (Sobek and Schuller 1978). A host of researchers have further refined this concept into practical methodology (e.g., Ziemklewicz 1990; Bradham and Caruccio 1990; Caruccio 1984).

The effective treatment of acid-producing mine spoils is generally achieved through several remedial concepts, some of which include lime treatment, encapsulation (burial), mixing or blending, bactericide treatment, and revegetation (wetlands and reforestation). Numerous studies have dealt with the lime treatment and encapsulation of acid producing rock and soil (e.g., Byerly 1990; Stiller 1982).

Reclamation efforts continue to improve the environmentally devastated strip-mined areas of the Cumberland Plateau. Ongoing research efforts will no doubt lead to further innovative techniques to mitigate the damage and help provide for a better environment.

Engineering Geology along Highways

The catastrophic effects of ignoring unstable geologic conditions are most readily visible along well-traveled highways. The cracking and sliding of highway embankments constructed on unstable ground, landslides blocking the highway, rockfalls pelting the roadway and obstructing traffic, and collapses of the roadway as a result of sinkhole development—all are examples of stability problems related to geology.

Over the last thirty years a new discipline of science has emerged, and its popularity is constantly growing: engineering geology, and highway engineering geology in particular. With the development of our nation's interstate highway program and its complex needs—a four-lane template, avoidance of steep grades, and relatively straight roads—has come an increased demand for geologists and engineering geologists in highway design and construction.

The geologic aspects of highway engineering design and construction are numerous and often unique, varying from geologic region to geologic region across not only the state of Tennessee but the entire country. From the need to share highway-related geotechnical problems and resulting solutions, a scientific organization called the Highway Geology Symposium has developed. The symposium enables geologists and engineers to bridge the gap of information from one region to another.

One of the chief geologic subjects of highway engineering has been the avoidance of landslides. Construction-caused landslides, of course, are the result of inappropriate design in areas with geological problems. Excavating a slope in the side of a hill composed of colluvium most likely will result in a landslide. The inappropriate location of a highway across karst terrain will usually result in a subsidence or collapse.

During the location phase of highway development, unstable areas are now bypassed if at all possible. In the past, unstable areas were either not identified or not acknowledged as unstable. Old landslides, unstable slope deposits, compressible and unstable soil, sinkhole plains, and unfavorable geologic structures are all now identified and avoided during the initial location of a roadway corridor.

For example, the location of the Pellissippi Parkway Extension into Blount County was altered in Knox County to avoid geologically unstable situations in the Ten Mile Creek and Keller Bend karst areas. The relocation of the roadway possibly eliminated numerous sinkhole collapse problems as well as extensive flooding in the Ten Mile Creek area.

Environmentally, the impact that is made by highway construction is quite evident. Earth and rock scars along ridges and mountains, erosion and siltation of streams (though mostly during construction), the altering of some stream channels, and the breakup of forest ecosystems are some of the impacts of new highways on the environment.

In their search for better ways of doing things, engineering geologists have applied the acid mine drainage problem and solutions to highway construction practices. Guidelines for handling excavated acid-producing materials have been developed for the Federal Highway Administration (Byerly 1990). Actual treatment of pyritic rock in conjunction with highway construction has resulted in refined encapsulation techniques using synthetic geomembranes (Fig. 51; Moore 1992). New highway corridors that have been constructed through terrain containing acid-producing rock include S.R. 165 in Monroe County, the Foothills Parkway in Blount and Sevier counties, and the U.S. 23 extension (future I-26) in Unicoi County. Although some difficulties were encountered during construction, especially along S.R. 165, the use of synthetic geomembranes has seemed successful in mitigating the acid drainage problem.

Acid drainage problems involve the excavation of rock formations containing significant amounts of iron sulfide minerals such as pyrite (fool's gold). Research into this problem has yielded new information regarding testing rock strata for acid drainage potential, treatment of

Fig. 51. Mitigating the possible acid drainage of pyritic rock involves the encapsulation (burying) of pyritic strata in geomembrane enclosures, which keeps the rock free from water and air.

acid-producing rock, and mitigation of environmental damage caused by acid drainage leachate (Byerly 1981; Moore 1992). In the last few years, the Tennessee Department of Transportation has begun to examine and test selected proposed highway routes for the potential for these problems in excavated rock.

Of particular interest is the Unicoi County U.S. 23 construction project in the mountainous region around Flag Pond, Tennessee. Over 10,000 rock samples have been tested from this highway route in order to locate potentially acid-producing rock. As of this writing, over 15,000 cubic yards of acid rock have been identified and encapsulated in order to prevent acid leachate problems from developing from the highway excavations. In addition, several tens of thousands of cubic yards of rock have been treated with agricultural lime to neutralize slightly acidic rock material not requiring encapsulation. Sampling and testing of the rock material before and during the construction project defines the location of the subject strata for easy and accurate removal and mitigation.

In the Unicoi County highway construction project, other environmental mitigation measures are also worthy of mention. Several millions of dollars were used to control erosion and siltation along the construction project through the use of extensive brush barriers, siltation fences (combined with hay bales), periodic seeding and mulching of bare ground slopes, and the installation of check dams and rip-rap ditches.

Another part of the Unicoi County project included the construction of lunker boxes (wooden boxes 18 inches high, 4 feet long, 2 feet deep, and open on one side) along a relocated section of South Indian Creek, a well-known trout stream. The relocated channel utilized the lunker boxes along the edge of the stream channel relocation in order to simulate the hollow spaces along the edge of a creek bank. The open space was placed facing the stream channel. In addition, hundreds of trees were planted along the bank to provide shade for the stream and its wildlife, and boulders were strategically placed in the stream channel to provide additional habitat in the stream and white water for other wildlife.

Among other environmental problems having a geologic origin is the erosion of soil from the weathered rock slopes. The use of serrated cutslopes in Unicoi County and elsewhere, however, has dramatically reduced the amount of silt and clay being eroded from the newly excavated cuts. Serrated cutslopes are the result of the construction of small, 2-to-3-foot wide and high, step-like benches (serrations) in the soil and weathered rock material. When seeded and mulched, these benches enable vegetation to establish itself without extensive sheet wash erosion.

Environmental problems, some of monumental proportions, are affecting us all, even in our everyday life situations. Landslides crack and break apart our highways, and sinkholes suddenly appear out of nowhere in our yards. The ravages of uncontrolled and unreclaimed stripmine areas are yet to be repaired. But scientists and engineers together are working on methods to correct or prevent these environmental problems. In your travels across Tennessee, you may notice some of the environmental problems related to geology. To recognize that such problems exist is half the battle in returning our planet to a more natural state of existence. The other half is for us to act responsibly and do something about the problems.

The Making of a Highway

In the early days a road often began as an animal path that through time had become a footpath. Continued use of footpaths and trails established such trails as accepted transportation routes connecting one village with another. When horses and wagons began to follow these trails, such use made them wider. Rain and snow, however, quickly rendered these travel routes quagmires, if not impassable. One of the earliest means of stabilizing these muddy roads was to use logs to bridge the unstable areas. This technique, called corduroy, involved laying logs and tree branches across the width of the road to provide a stable roadbed in the muddy areas.

One of the earliest highways constructed in the United States was the Boston Post Road, which by 1673 connected Boston, Massachusetts, with New York City. After the Revolutionary War, the debt-burdened state governments chartered private turnpike companies to build roads and charge tolls for their use. One of the early turnpikes was a 62-mile-long road built in Pennsylvania between Philadelphia and Lancaster from 1793 to 1796. That particular road, 21 feet wide, was constructed of gravel on top of a roadbed of wood, stone, and gravel (Tyler 1957: 11).

In 1775, Daniel Boone and about thirty other men staked out the Wilderness Trail from Cumberland Gap into Kentucky. This trail provided access into the Kentucky region from Tennessee and Virginia (see Side Trip 1 to Cumberland Gap National Historic Park).

In 1811 construction began on one of the first personally financed roads in the United States: the Cumberland Road, stretching from

Cumberland, Maryland, northwest to the Ohio River at what is now Wheeling, West Virginia. The Cumberland Road was 30 feet wide, the central 20-foot section paved with several layers of stone. Its 66-foot-wide right-of-way was cleared and had drainage ditches along the roadbed (Rae 1971: 18).

By 1872 the first brick pavement was constructed in Charleston, West Virginia. In 1893 the Office of Road Inquiry, forerunner of the present-day Federal Highway Administration, was established within the Federal Department of Agriculture (Tyler 1957: 15).

During Reconstruction (1865–1900) after the Civil War, few highways were built, as most people depended on the rapidly developing railroads for long-distance transportation. By 1900, however, the popularity of a new invention—the horseless carriage—provided an impetus for road building. The public demand for roads on which to drive their new automobiles increased with the number of cars in the United States, which grew from 8,000 in 1900 to 20 million by 1925 (Tyler 1957: 15).

In 1916, Congress passed a law that began a policy of providing Federal funds to aid the states in the construction of roads (Tyler 1957: 15–16). Each state was required to establish a state highway department to receive the Federal funds. The major emphasis by the individual states was to provide paved roads from rural areas to city markets in order for farmers to transport their produce and livestock. With this improvement in roads came the development of the trucking industry in the United States.

In 1919 a momentous event occurred that was to affect the motoring public even to this day: in the state of Oregon, the first gasoline tax was adopted (Rae 1971: 62). It is through the revenues of the Federal and state gasoline taxes that highway improvements are funded today.

With the rapid increase of automobile ownership in the United States during the 1930s and 1940s came the necessity for upgrading the road system. Automobile registrations increased from approximately 8 million in 1920 to over 40 million by 1950 (Rae 1971: 50).

When the United States entered World War II, gasoline rationing and a limited supply of construction materials brought both the automobile industry and road construction to a major slowdown in U.S. history (Tyler 1957: 17–18). After the end of the war, however, road-building programs were renewed, providing not only better roads but safer ones.

The construction of roadways in the early days involved simply stabilizing the existing path or trail. With the appearance of the automobile, these transportation routes quickly became rutted.

Accordingly, crushed stone became widely used to provide a stable roadbed. Even today, gravel roads are still common, especially in rural areas. Asphalt, however, quickly became the material of choice to stabilize the roadbed gravel and provide a smooth riding surface.

With 15 million carts, wagons, and carriages on the streets and roads of the United States in 1878, the mostly dirt and gravel rural roads caused great difficulties with dust (Rae 1971: 27). Early attempts to remedy the situation with these dusty roads involved the application of oil and tar to settle the dust, a technique first used on some roads of Los Angeles County, California, in 1898 (TDOT 1976: 67). Although brick and asphalt paving were being used on city streets during the early 1870s, these materials were much too expensive to be used for the many thousands of miles of rural roads (Rae 1971: 27; FHWA 1976: 67). Choices for roadbed surfaces expanded in 1893 with the introduction of concrete for that purpose in Bellefontaine, Ohio (Rae 1971: 27).

By the early 1900s the demand for surfaced roads and, indeed, better highways increased to the point that the Federal Government through the Federal Aid Program of 1916 provided the funds necessary for paving many thousands of miles of highways in the United States. Although by 1904 only 18 miles of American roads were asphalt, 16 miles of which were in Ohio (Rae 1971: 32–33), by 1957 paving had been done on 854,000 miles of rural roads and city streets, 450,000 miles of them asphalt (Tyler 1957: 26).

As technology developed, the roads were straightened or new routes were built to provide more direct routes between cities. In the early 1920s and 1930s little planning was carried out in the development of these new roads. In those days, few people had formal training or education—let alone specialization—in the design and construction of highways.

Early Roads in Tennessee

The road through Cumberland Gap, part of what was known as the Great War Path, was used as early as 1750 by Dr. Thomas Walker, who gave Cumberland Gap its name (TSHD [Tennessee State Highway Dept.] 1959: 2). This same road, later called the Wilderness Road, was continued into Kentucky by Daniel Boone.

Another early road was the one from Burke County, North Carolina, to Jonesborough (capital of the State of Franklin from 1784 to 1788), along which many settlers came into what is now upper East Tennessee.

In 1780 a path was at least partially constructed from Watauga in upper East Tennessee to Nashborough (later Nashville) in Middle Tennessee. In 1787 the North Carolina Assembly made provisions for a new road to be opened by the militia from Campbells Station in Knox County to Nashville in order to aid the settlement of the new territory (TSHD 1959: 2–3).

As the pioneers pushed westward, more new trails and roads were opened. By 1817 approximately 1,500 miles of roads were established in Tennessee, many of which had been developed from Indian and buffalo trails (TSHD 1959: 11). The Great National Road, commissioned in 1826 to connect Washington, D.C., to New Orleans, was to pass through the center of Tennessee, then westward through Memphis (TSHD 1959: 12).

In Tennessee, as in the rest of the United States, the period of Reconstruction was not a time of progress in road building. A notable exception was the construction in 1889 of the first public road system in Tennessee (TSHD 1959: 14). By 1900, the automobile craze had hit: of the approximately 8,000 autos registered in the United States, 40 were in Tennessee (TSHD 1959: 16). To meet the demand for new roads in the state, the Tennessee General Assembly established the State Highway Department in 1915 for the control of state highway construction and maintenance. By 1919 the state highway system consisted of about 4,000 miles of important routes of statewide interest (TSHD 1959: 27).

The Great Depression affected the Tennessee road-building budget. In 1932 expenditures were reduced by two-thirds—from $33,105,324.22 in 1930 to $10,435,881.55 in 1932 (TSHD 1959: 42). This reduction of funds limited the amount of road improvements that could be made by the State Highway Department. In fact, it was not until after World War II that significant improvements were made to the state's road system.

Only in 1947 did the state budget for roads again exceed $10 million ($17,447,808.18 was expended that year). During the late 1940s and early 1950s, increased automobile traffic necessitated new and improved roads for the motoring public, and the State Highway Department evolved into a multidivision operation.

In 1923 there were 173,366 vehicles registered in Tennessee; by 1955 there were 1,104,650. The annual state highway budget increased comparably: from $3,790,834.91 in 1923 to $51,725,253.21 in 1955 (TSHD 1959: 96–98). (For comparison, the 1991–92 total budget for Tennessee highways was just under $1 billion.) As a result of the dramatic increase in autos during the late 1940s and early 1950s, a significant development in the nation's highway system was enacted in 1956.

The Interstate System

Over the years, the transportation needs of Tennessee and other states have grown from two-lane roads to four-lane highways and finally to an interconnecting system of high-speed, controlled-access highways called interstate highways. In 1956 the Congress of the United States established the Highway Act of 1956, which set up a 41,000-mile system of superhighways called the Interstate and Defense Highway System to connect every major city in the United States (Pack 1964: 12). Tennessee was allocated an original 1,047.6 miles of interstate highway mandated to be complete by 1972; the first section of interstate in Tennessee was a 1.8-mile stretch of Interstate 65 in Giles County that was completed on December 17, 1958 (Moulton 1960: 22).

In May 1957 it was estimated that of the total interstate mileage allocated 862.9 miles would be in rural areas and 121.6 in urban areas (the total, 984.5, is minus some interstate loops and connectors). The cost estimate was $1,076,299,000. In June 1962, however, the Department of Highways (as it was called at that time) was able to reduce the estimate by $245 million by using new locations for the interstate and new technological methods for the design. (The Department of Highways became the Tennessee Department of Transportation in 1972 during the administration of Governor Winfield Dunn.)

Since 1958, when the first section of interstate was opened to the public, nearly 1,000 miles of interstate highways have been built in Tennessee, including complete sections of I-40, I-75, I-81, I-24, and I-65. Loops and spurs are continually being added to the system as the demand for expansion of our highways continues to grow.

Soon, an I-840 loop south of Nashville will be completed, connecting onto I-40 at Dickson in the west and Lebanon in the east. In addition, the construction of a proposed Interstate 26 in Unicoi County will link I-81 in Sullivan County with Asheville, North Carolina, where I-26 is being continued toward the Tennessee state line.

The Development of I-40

I-40, by which you are taking your geologic trip, has a long and complicated history that goes back to the 1950s. By February 14, 1956, the first section of I-40 was let to contract for design or construction. In Shelby County, this first stretch of road was to run from Nonconnah Creek to Hindman Ferry Road at the Wolf River. Davidson County became the next county to have a section of I-40 placed under contract, when the design contract was let on March 4, 1956 (Leech 1956: 80).

By 1957 other portions of I-40 were under contract (preliminary design, right-of-way acquisition, or construction) in Knox, Shelby, Davidson, Roane, Haywood, Madison, Jefferson, and Cocke counties. By 1958 additional contracts were let in Loudon, Smith, Putnam, Cumberland, Humphreys, Hickman, and Sevier (Leech 1958: 82–83). Among the portions of I-40 completed by 1958 was the section in Knoxville from the I-40/I-75 interchange to 17th Street (Leech 1958: 16), later dubbed "Malfunction Junction" by the motoring public.

Between July 1, 1958, and June 30, 1960, 152 miles of interstate highway were placed under construction (Moulton 1960: 21). By 1960 additional design and construction contracts were let in Wilson, Dickson, Williamson, Henderson, Benton, and Haywood counties (Moulton 1960: 76–79).

During the period between July 1, 1962, and June 30, 1964, design and construction contracts for I-40 were let in not only the counties already mentioned but also in Fayette, Carroll, and Decatur counties (Pack 1964: 74–78). By that point 252 miles of the total interstate system had been completed in Tennessee and an additional 348 miles were under construction (Pack 1964: 12).

By the late 1960s most sections of I-40 were either completed and open to traffic or under construction. Problems, however, had begun to surface. Many were related to environmental concerns, such as the landslides on sections of the interstate over Rockwood Mountain in Roane County and the impacts on Overton Park in Memphis, matters that will be discussed later in this chapter.

Phases of Highway Construction

You may wonder why it takes such a long time to build a road like I-40. What most people see, however, is only the construction phase of highway development, and that is actually the last main phase in the development of a highway. A number of steps are required before the highway actually goes into the construction phase. Planning, location studies, and engineering design are all necessary before the highway is actually built. Once into the construction phase, the highway may take from one to two years—and in some cases, much longer—depending on site conditions, materials, weather patterns, and changes in the roadway design.

Research and Planning

The initial phase in the development of a highway is research and planning. This involves traffic counts and user patterns along subject highway routes. Research from these counts and pattern uses will guide the design engineer in formulating a design concept for the highway. Information gathered in this planning stage is also used for new highway locations (where a highway facility may not be in existence).

Traffic counts are first gathered for hourly traffic flow, which culminates in an average daily traffic count (ADT). These count figures are used to decide whether or not to build a two-, four-, or six-lane facility or to justify the widening of an existing two-lane road. This information is also used to design improvements for intersections on existing roadways.

Funding

Once a conclusion is made based on the research data, the highway project is then subject for funding. Administrative decisions are made on the type of funding to be acquired. This may involve local, state, and federal funds, with complex funding formulas to be worked through. At this point, the order of priorities may be influenced not only by such matters as safety concerns and traffic demands but also by political influence. Once appropriations are made, the highway project enters the conceptual design stage. This complex process involves studies on environmental conditions that may be affected by the roadway project.

Location Studies

The role of the Geotechnical Office in developing the highway plans begins in the location phase. The selection of the site for the proposed roadway involves strategic information provided by technical experts in several fields, including geology. Location studies are undertaken to find the best route possible for the new road (Fig. 52). General geometry of the highway is formulated against the topographic features of the area. These features include the configuration of the land surface (e.g., ridges, mountains, and valleys), river and small stream crossings, population density, urban, commercial and industrial development areas, and the location of existing highways. Cemeteries, schools, and hospitals are avoided if possible, as are areas of high-dollar real estate and obvious geological problems.

Fig. 52. Locations for new highways are studied for environmental impact and proper design.

Initially, a corridor location is selected. It may contain two or three alternates. Additional engineering studies are made at this point to evaluate the best possible route location. Citizen input in the form of a preliminary public hearing is also incorporated into the process. Once a final location for the route is decided upon, then a series of environmental concerns are studied and addressed. Items of environmental concern include plant and animal life, surface and groundwater conditions, unusual geologic conditions such as landslides, sinkholes, and toxic rock, as well as archaeological and historic sites.

Some of the more significant environmental considerations associated with highway construction include landscape scars, erosion and siltation (mainly during construction), the cutting of trees, effects of acid-producing rock, and impacts on wildlife and streams. Years ago these topics were not generally considered during the development of our highway system. During the last twenty years, however, a great surge in the public awareness on environmental concerns and issues has prompted government agencies to adopt new approaches to highway development, to develop new design and construction techniques, and to implement appropriate mitigation measures regarding environmental issues.

The Overton Park controversy. From 1956 to 1970 numerous meetings were held in Memphis concerning the route of I-40 through Overton Park in downtown Memphis. The 342-acre park is a tract of land purchased by the City of Memphis from Overton Lee in 1901. The people of Memphis voted in 1902 to name the park in honor of a city founder and benefactor, Judge John Overton.

Overton Park is of great environmental significance. Though in an urban area, it has a wealth of flora and fauna. The park contains, for example, over 75 varieties of trees and numerous wildflowers. Moreover, it not only has a significant resident population of birds but also serves as an important resting area for birds in their heavy migration along the Mississippi River flyway. In addition, a number of mammals, such as rabbits, squirrels, raccoons, opossums, and mice are found in the park (TDOT [Tennessee Dept. of Transportation] 1977: 3–16).

Because of its convenient location in Memphis, Overton Park provides many thousands of people with natural areas and recreational opportunities for their enjoyment. The park includes a municipal zoo, a nine-hole golf course, hiking trails, and a small lake.

In December 1969 a local group called Citizens to Preserve Overton Park, Inc., took legal action in the U.S. District Court to enjoin the U.S. Secretary of Transportation from releasing Federal funds to the Tennessee Department of Highways for the construction of I-40 through Overton Park. The district court found that the Department of Highways had made decisions regarding the I-40 project through Overton Park that were in compliance with the 1968 Federal-Aid Highway Act. The citizens group then appealed all the way to the Supreme Court, which decided that the case should be sent back to the U.S. District Court. In January 1972 the U.S. District Court determined that all alternatives had not been adequately analyzed and directed the Federal Highway Administration and the State of Tennessee to provide alternative designs and locations for the route of I-40.

Other alternatives involved the disruption of local neighborhoods with alternate routes and the construction of a partially buried tunnel through the park. Citizens continued to object to the route through the park. Eventually it was decided to route I-40 around the north side of Memphis with I-240. This is the route that I-40 takes today.

The Overton Park controversy is a good example of how a group of concerned local citizens can alter the location of a highway project. Indeed, the involvement of concerned people in public works projects can produce more beneficial results—and certainly ones that are better accepted. It is important, however, that the initial involvement occur at early stages of the highway planning process.

Design

Basic roadway design begins once a route location is selected. At first, general conceptual design is developed. This phase includes the development of the general geometry of the road alignment, typical roadway width, topographic information (such as contours of the land surface and location of houses, existing roads, and utilities) and functional layout of the proposed roadway.

Upon completion and approval of the functional design, a more detailed and complex design is developed. This phase includes the establishment of a centerline profile of the roadway along the length of the project. A centerline profile is a detailed survey of the land surface along the center of the proposed roadway. Land-surface profiles of cross-sections are also surveyed along short sections perpendicular to the centerline profile. These cross-sections are spaced at 50-foot intervals along the length of the centerline.

At this stage, the proposed roadway template is drawn on each of the cross-sections. An appropriate roadway cut-and-fill slope design, based on geotechnical analysis, is added to the cross-sections during this phase. A *roadway cut* is an excavation into or through a hill or ridge where a road is to be located. The term *fill* refers to a place where the material excavated from a hill (cut section) is placed in a valley or other low area to build up the roadbed; a fill is also known as an embankment. In essence, when a road is designed and constructed, the concept is to excavate material from the high places in the land surface and put it in the low areas so as to use all the material excavated as part of the roadbed. Accordingly, this type of construction is called a cut-and-fill roadway.

During the design phase, geotechnical input is generated in three areas: roadway, structures (bridges and retaining walls), and special geotechnical considerations. Routine auger and core drilling determine soil depths, rock types, and rock location (Fig. 53). Soil samples are taken to determine classification, proctor density, and moisture content and in some instances stability parameters (cohesion and internal angle of friction). Rock samples are taken to determine rock type, chemical properties, and strength characteristics.

Fig. 53. Geotechnical studies of new roadway locations often involve drilling and sampling the soil and rock.

Geotechnical input into the highway design process also involves the determination of appropriate design concepts and formulation of the required design parameters. The angle on which to make a resulting slope from an excavation (cutslope) is determined from the soil properties and/or rock characteristics.

Construction of bridges and retaining walls must rely on adequate information about their foundations. The geotechnical investigation of a proposed structure site usually involves drilling where the proposed footings are to be located, taking samples of soil and rock to determine strength parameters, and interpretation of the subsurface conditions to develop appropriate design. Through a subsurface investigation, unstable soil and cavity-riddled rock can be identified and adjustments made to either avoid or accommodate the instabilities in the foundation design.

Identification of geologically unstable areas along proposed roadway routes usually requires special design attention. When such features as ancient landslides and weak soil or rock conditions exist, then special design concepts using rock buttresses (large retaining structures that depend on the force of gravity) or conventional retaining walls may be used. Less conventional restraining methods have been used along I-40 on Rockwood Mountain, including gabion wire baskets and reinforced earth walls (both are retaining wall systems).

In some areas, rock cutslopes may require special attention because of structural weaknesses in the rock mass. Fractures (joints and cleavage), fault planes, and bedding planes all provide discontinuities along which a rock mass may experience instability. Rock bolts, wire meshing, trimming and scaling, catchment fences, ditches, berms, and horizontal drains have all been used to correct these instabilities. This is especially evident along some East Tennessee stretches of I-40 in Cocke County.

During the design phase, additional information concerning property lines and owners, detailed locations of utilities (e.g., water, gas, power) and buildings are added to the plans. With this information detailed and drawn on the plans, then the appropriate right-of-way boundaries can be established. Once right-of-way approval is obtained (from public hearings and engineering requirements), then the purchase of the needed land is initiated.

Appraisers first inspect and evaluate each parcel to be taken and make a report with the appraised value established. Next the buyers representing the state go to each property owner and make the necessary offers for the desired parcel. This process sometimes involves counter offers and legal assistance before the property is finally acquired.

During the right-of-way acquisition process, the roadway plans are further developed and the final design of bridges and retaining walls determined. During the design stage, the Geotechnical Office conducts the necessary geotechnical studies, which include drilling investigations (including sampling of soil and rock for testing purposes), field mapping, analysis of field data, and calculations for design. Most of the geotechnical design involves establishing the appropriate angle of slope on which to excavate a cutslope or construct an embankment. A typical soil cutslope may require an angle of approximately 26.6°, which is commonly referred to as a 2:1 ratio (that is, a slope angle constructed on two horizontal units of measure to one vertical unit of measure). A typical rock cutslope, however, may require a vertical slope or even a slope on a 63.4° angle, called a 0.5:1 slope.

Roadway embankments (fills) are also designed to be constructed on a specified slope angle that the particular fill material used can be stacked on without becoming unstable. The stability of the material depends upon the properties of the soil and rock, such as grain size, moisture content, clay content, and density. Fill slopes are usually designed and constructed on angles that are 26.6° or flatter.

The results of these studies include the establishment of detailed soil and rock profiles along the centerline of the proposed roadway, rock and soil slope design (for cut-and-fill intervals), the location and treatment of unstable conditions (such as soft or wet ground, sinkholes, acid-producing rock, or landslide areas), and the formulation of foundation information and design recommendations for bridges and retaining walls. Retaining walls are engineered walls, usually constructed of steel-reinforced concrete, which keep soil or rock from sliding into the roadway in a cut or fill section. Retaining walls are normally used in urban areas where land space is restricted. Good places to see retaining walls are on I-40 in Knox County (Road Log, Miles 387.0–376.0), Davidson County (Miles 213.5–204.1), and Shelby County (Miles 10.8–0.0, including I-240 North). In addition, the recently constructed I-240 loop in Nashville uses retaining walls extensively to minimize the impact of the highway on residential areas.

Once the geotechnical information is added to the roadway plans, then special attention is given to construction details involving geotechnical expertise. These include type and quality of rock for rock drainage pads, unstable soil or rock conditions, remedial direction for landslides and embankment failures, and special treatment for unusual geological problems such as sinkholes and caves, acid-producing rock, mineral deposit evaluation and extraction, and landfill stability and leachate problems.

Developing the construction plans also involves setting construction limits, special permits, easements, and rights-of-way; specific construction details including items required, quantities to be excavated, placed, bought and/or removed; and the addition of the final cross-section.

Bidding and Review

The proposed highway construction documents are then advertised for bidding purposes. Once bids are received from the contractors, then extensive review of all items in each bid is required before awarding the contract to the lowest approved bid. After the contract has been awarded, special attention is again required for reviewing the proposed plans for discrepancies and changes. Once construction starts, careful monitoring of the project is conducted by the office of the TDOT Project Engineer. Surveying, inspection, quality control, and testing are all required in the supervision of a project by the TDOT Construction Office.

Construction

During the construction phase of road building, reshaping of the land surface occurs. It is usually necessary to change the topography, cutting down hills and filling in valleys, in order to build a new road (Fig. 54). Powerful machinery—bulldozers, front-end loaders, pans, and dump trucks, for example—are usually required for moving the soil and rock encountered during road construction.

Knowing the geologic conditions along a proposed roadway enables the design engineer not only to make appropriate design decisions but also to calculate how much soil and rock will be removed during construction. It is critical to know the amount of soil and rock to be removed on a project in order to plan the appropriate construction techniques and to arrange for the necessary equipment. During the construction phase the geologic conditions (e.g., types of rock, rock structures, and soil thickness) are actually encountered and exposed.

Generally, construction of the road begins in embankment areas, which are prepared by constructing or placing pipe culverts in the low places for surface drainage to flow through. Next (or sometimes

Fig. 54. Most highway construction involves cut-and-fill operations where high spots are excavated and deposited into the valleys; view is of I-26 construction in Unicoi County.

occurring at the same time) is the excavation of cut sections where soil and rock are removed from the proposed roadway location (Fig. 55).

The soil is usually removed by bulldozers and/or earthmoving pans. A pan is a large, four-wheeled motorized piece of equipment with a large scoop (pan) between the front and rear wheels. As this earthmoving machine rolls over the soil, the pan device is dragged along, scooping up soil and weathered rock, sometimes as much as 35 cubic yards at a time. Rock that is to be removed is usually drilled and blasted (shot) into smaller fragments. Then a front-end loader scoops up the rock fragments and dumps them into a large dump truck to be hauled away.

After the excavated soil and rock are transported by truck to the embankment areas, the roadway fill sections are constructed. Embankments (fills) are made from the excavated materials, usually soil, rock, or a mixture of the two (Fig. 56). The soil is spread in layers of about 8 to 10 inches at a time, eventually building up the roadbed to the required elevation.

As the soil is placed in the embankment areas, it is compacted to squeeze out the air between the particles of soil. Steel rollers shaped like barrels with steel rods projecting out are pulled (rolled) across the soil. The steel rods push into the freshly layered soil, compacting it. This process is very important, for it keeps the soil embankments from settling—and perhaps even sliding—once construction is over.

Fig. 55. Cut slopes through rock are often cut smooth by a process of pre-splitting the rock face during construction.

Fig. 56. Fills (embankments) are constructed where material excavated from hills is placed in low areas or valleys in order to build up the roadbed.

When rock material is used to build an embankment, however, a compacting device is not usually employed. The rock fragments are kept to diameters of less than 3 feet (typically 8 to 18 inches) and placed in layers that are 18 inches to 3 feet thick. This excavation process usually takes a number of months, depending on the size of the roadway project (the length of the particular stretch of road and the quantity of material to be excavated, for example). In the case of I-40, construction required the excavation of several hundred million cubic yards of soil and rock.

Once the roadbed is constructed to the elevation that the roadway plans require, then surfacing the roadbed generally follows. Surfacing is the process of placing a layer of material (e.g., gravel, asphalt, or concrete) on the roadbed so that vehicles can drive on it. The type of road surface chosen depends mainly upon the quantity and the nature of the traffic that will be using the road. High-volume interstate traffic with numerous trucks, particularly in cities, usually requires concrete pavement, which is more substantial for resisting such traffic loads. In rural areas, which have less traffic than urban ones, interstate surfaces are usually asphalt. Other factors, however, enter in: for example, cost and availability of the material (asphalt is generally cheaper than concrete). Whatever the choice of surfacing material, a layer of crushed stone, 6 to 10 inches thick, is placed on top of the subgrade bed to provide a stable layer for the road pavement to be placed upon.

The TDOT Materials and Tests Division conducts periodic tests during construction on all materials used and/or purchased for the project. Concrete is tested for strength and correct chemical mixture, roadway stone is tested for quality and appropriate gradation, steel reinforcing bars are tested for correct thickness and steel strength, and the metal guardrail is tested for the required galvanizing thickness as well as strength.

When the asphalt pavement is placed, the asphalt is tested to see if the correct mixture of asphalt and stone is used; even the gradation of the asphalt-stone mixture is measured. Once the paving process is complete, the white and yellow paints used to stripe the lanes of traffic are tested for correct chemical composition. Because the paints also contain tiny glass beads (smaller than a grain of sand) for reflective purposes, they must also be checked to make sure these beads are present in the correct number and size.

Geotechnical input is also required during the construction of roadway projects. One of the main areas of geotechnical responsibility concerns the approval of foundations for such things as bridge and retaining wall footings, rock buttresses, and the type and quality of rock used for rock pads and rock buttresses. Sometimes, special geotechnical problems arise during construction such as landslides, cave-ins, and other areas of instability that require geotechnical expertise for correction.

I-40 landslide area on Rockwood Mountain. During the late 1960s and early 1970s, the completion of I-40 across Tennessee was hindered by massive landslides that began developing in East Tennessee along the graded portions of the early I-40 route.

The route crosses from the Valley and Ridge province to the Cumberland Plateau province in rural Roane County just northwest of the town of Rockwood. The border between the two provinces is an escarpment about 1,000 feet high. Construction of a highway across the escarpment was the sort of challenging project that demanded geotechnical review to locate and design the roadway appropriately. Unfortunately, geotechnical involvement was not a part of the highway planning process at the time the road in Roane County was located and designed.

For many years landslides have plagued all types of roadways crossing the Cumberland Escarpment. Most of these landslide failures have resulted from the disturbance of the relatively tenuous state of equilibrium that exists along the interface between the colluvium (a mixture of gravity-deposited boulders and soil, usually along the base of high escarpments) that drapes the escarpment slope and the underlying residual soils (most notably, the residuum of the Pennington Shale). The colluvium is very porous and transmits large quantities of groundwater near the surface. The Pennington Formation contains shale mixed with clay minerals that absorb water and become very plastic and slick; the result is a loss of shear strength. At best, the natural condition is only marginally stable and likely to permit movement of the colluvium. When highway embankments are constructed over this material, the added weight of the embankment on the colluvium increases the tendency of the colluvium to lose resistance to shear and slide along the slick Pennington clay surface. The high rate of precipitation along the escarpment, coupled with drainage problems, exacerbates the breakdown and loss of shear strength of the shale roadway embankments, producing additional landslide problems along the plateau highways.

Landslide problems along this five-mile section of I-40 (Miles 346.5–341.5) began in January 1968, with more than twenty major landslides and embankment failures developing along the eastbound lanes during that year (Royster 1973: 255–56). Over the next several years additional slides involved both the lower eastbound and upper westbound lanes. Roadway construction soon ground to a halt as geotechnical investigations and remedial design were required.

It was at this point, in 1971 and 1972 (during the administration of Governor Winfield Dunn) that the Tennessee Department of Transportation wisely established the Division of Soils and Geological Engineering. This geotechnical division was created not only to investigate and design remedial concepts for landslide repair but also to function as an integral part of the Transportation Department's overall program.

After intensive geotechnical investigations a number of remedial designs were developed. Various forms of remedial measures were employed, including partial to complete removal of the landslide material, partial removal combined with restraint of the landslide, total restraint through extensive use of rock buttresses (gravity structures used to retain large masses of earth), and drainage, by means of ditches, wells, and horizontal drains (horizontally drilled wells that allow the groundwater to drain out freely by gravity flow) to stabilize most of the landslides (Royster 1973, 1977, 1979).

The various methods of restraint included rock buttresses, gabion walls, and reinforced earth walls. Rock buttresses are free-draining gravity structures consisting primarily of large blocks of nondegradable sandstone or limestone. Movement of the unstable material is prevented by restraint (by the weight of the rock buttress) while at the same time allowing large quantities of groundwater associated with these slides to percolate or seep through the rock. Obviously, the foundation of one of these large rock buttresses needs to be on material that can withstand the weight (usually in-place bedrock).

The numerous rock buttresses along this section of I-40 are characterized by their large masses of clean "chunk" rock (pieces of gray limestone the size of small boulders). These structures are most easily seen from the eastbound lanes (as you travel down the mountain) where these buttresses have been constructed between the westbound and eastbound lanes.

Gabion walls, another type of restraint, are free-draining, heavy, monolithic gravity structures consisting of zinc-coated wire mesh baskets (gabions) filled with coarse, nondegradable rock (Royster 1973). Gabion walls were used along sections of the Rockwood Mountain traverse because they were relatively low in cost, they provided flexibility needed to cope with the variable site conditions, and they provided excellent permeability needed for the groundwater drainage. Five gabion walls were constructed, all along the westbound lanes.

A reinforced earth wall was used to correct a major embankment failure along the eastbound lanes of I-40 near Mile 342.2 (Trolinger 1975). A reinforced earth wall is a structure composed of horizontal strips of galvanized metal attached on one end to concrete face panels. A layer of granular backfill (usually sand or clean gravel) is placed over each layer of the strips, usually requiring twenty to thirty vertical layers of metal strips and backfill (depending on the designed wall height). The entire mass exerts a vertical force on the metal strips that develops into a resisting horizontal pullout force. The earth mass then acts singularly as a retaining wall, resisting the sliding force of the landslide. The concrete panels serve only to keep the granular backfill from spilling out of the structure.

As you can see, it required a variety of geotechnical designs to combat the landslide problem along Rockwood Mountain. Other more geologically suitable routes, of course, could have been used for the I-40 location in the first place. However, recommendations from TDOT's geologists and soils engineers were apparently not well received by the late-1960s state administration.

According to TDOT records, over $20 million was spent to stabilize this five-mile stretch of highway. Even today, stability problems exist, though they are constantly monitored by TDOT personnel. This particular stretch of highway should remind us all—politicians as well as the general public—that the costs of manipulating our environment can be quite high when certain aspects of the environment are either not well understood or simply ignored.

Maintenance

Maintenance problems, too, often require geotechnical insight and conceptualization. Some of the particular problems requiring geotechnical expertise to remedy include landslides, rockfalls, embankment failures, sinkhole collapse, and flooding.

I-40 rock slide area. Rock slides have been a major maintenance problem along I-40 near the North Carolina–Tennessee state line. This easternmost section of I-40 in Tennessee (Miles 451.8–448.2), in mountainous Cocke County, has been plagued with slides occurring in two modes: single boulders and wedge failures (Fig. 57). I-40 was repeatedly closed for up to two weeks at a time by wedge-failure landslides. Although, luckily, no one was killed in these landslides, the debris and rock slides did cause property damage and minor injuries. The massive construction project to repair this three-mile section of I-40 in the mid-1980s cost over $9 million.

Mitigation of the wedge-failure landslides first involved design work in the late 1970s and early 1980s. TDOT has implemented two main methods to solve, or at least diminish, these problems: stabilization and protection (Moore 1986: 448–49). Three types of stabilization methods—drainage, removal, and restraint—have been used, either separately or in combination, to stabilize these rock slope masses so that they do not move downslope.

Drainage methods usually involve dewatering a rock mass by means of horizontal drains; these horizontally drilled holes are left open in the

Fig. 57. Unstable rock slopes such as this on I-40 near Log Mile 449.2 can cause landslides that block the roadway. TDOT photo by George Hornal.

rock and drain by gravity flow. Pumps (not unlike the ones used with the familiar household water wells) and diversion ditches have also been effective.

In 1979, 24,650 linear feet of horizontal drains were installed along this section of I-40, just east of the Hartford interchange (Royster 1979; Moore 1986: 450–51). The horizontal drains proved successful in stabilizing a very large wedge-failure landslide (over 2 million cubic yards in volume) now referred to as the Hartford Slide (see Road Log, Mile 448.0).

Removal techniques include normal mass grading operations, controlled blasting techniques, and cleaning out debris at the bottom (toe) of a slide. All of these procedures were used successfully on I-40 through the Pigeon River Gorge. Most noteworthy are the total removal operations of the Waterville Slide (across the river from I-40 at Mile 450.2) and the oriented pre-split removals implemented at the wedge-failure slide area along the westbound lane at Mile 449.2. The oriented pre-split procedure involves the removal of existing wedges of unstable rock by developing a pre-split blasting face that is oriented to intercept stable planes of bedrock, removing the potential hazard (Moore 1988: 241) (Fig. 58). The pre-split rock faces are those odd-looking ledges of rock with evenly spaced parallel vertical lines that are drill holes (Fig. 59).

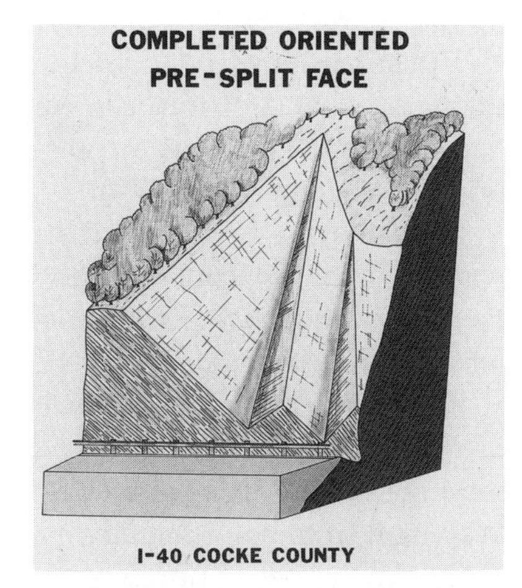

Fig. 58. These schematic drawings illustrate a before-and-after situation along I-40 at Log Mile 449.2, where oriented pre-split techniques were used to remedy a wedge-failure–prone cut slope.

Fig. 59. Using a technique called "oriented pre-splitting," the Tennessee D.O.T. changed this landslide-prone cut slope into a clean and safe rock slope (I-40 Log Mile 449.2). TDOT photo by George Hornal.

The restraint of rock wedges has involved the use of rock anchors and post-tensioned rock bolts (essentially large steel bolts, 1 inch in diameter) that are fully cemented (grouted) into a rock mass and tensioned so as to exert a compressive force, keeping the rock in its original position (Mile 451.4).

Protection methods include structures designed to protect the roadway and motoring public from rockfall hazards. Among these structures are wire meshing, catchment fences, and catchment ditches, all of which were used along this section of I-40.

Wire meshing—either wire rope nets or fence meshing (chain-like or gabion)—is draped over the precipitous rock slope (Fig. 60). The meshing holds a rock boulder against the existing rock slope as it falls, keeping it from bouncing down and out onto the roadway.

Catchment fences (for example, at Mile 451.4) form a flexible barrier that protects the roadway from falling rocks, dissipating the energy of those that fall rapidly. The rock catchment fences are used where ditch width is narrow or restricted. The height of the fence is used to compensate for the lack of ditch width.

I-40 sinkhole area. A stretch of I-40 in Loudon County (Miles 367.0–365.2) crosses an area underlain by cavernous rock strata of the Knox Group. Since its construction this section of I-40 has experienced several sinkhole collapses, some of which occurred near the roadway shoulder (Fig. 61).

Geotechnical treatment of these collapse features usually involves a method of bridging the sinkholes to provide adequate stability (Moore 1976, 1984, and 1988). Among these methods are chunk rock backfill, rock pads, concrete pads and concrete structures (such as bridges or walls), grouting, geofabrics, gabions, and lime stabilization. Other remedial techniques implemented in areas where highways cross karst include the use of paved ditches (Moore 1987: 118), artificial collapse (blasting and excavation in rock and dynamic compaction in soils), doline cleanout and protection, curbing for embankments, and overflow channels (Moore 1988: 137–43).

Along this section of I-40, chunk rock backfill was used along with paved ditching to correct the sinkhole collapse features. The collapse areas were initially excavated to remove the failed soil down to bedrock. Afterwards, the resulting holes were backfilled with clean chunks of limestone. When the backfill stone approached the roadway

Fig. 60. High-strength wire mesh is draped over rock-fall–prone cut slopes to prevent the falling rocks from reaching the traffic lanes.

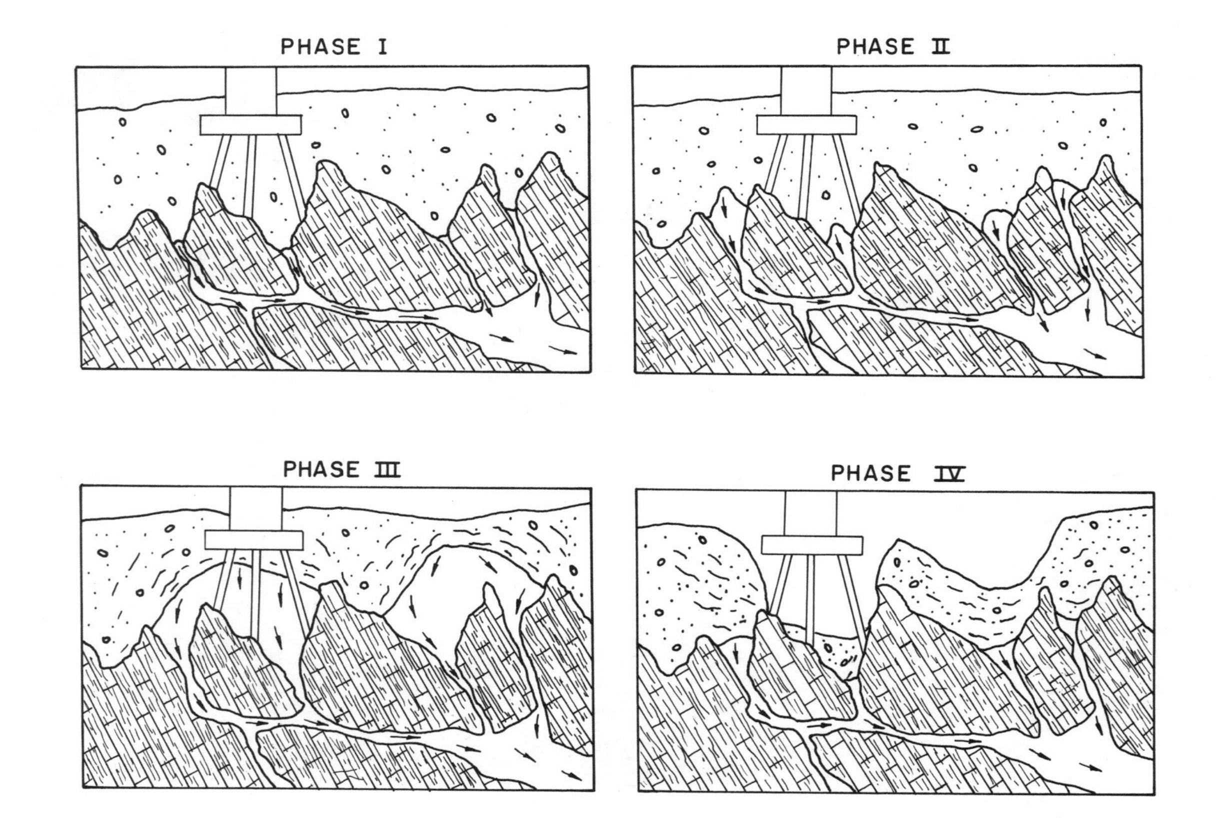

Fig. 61. This series of schematic drawings represents the development of a collapse failure of a bridge foundation constructed in karst terrain. Phase I. Pier footing, steel piles, soil, cavernous rock. Phase II. Pier footing, steel piles, soil, cavernous rock,

subgrade base vertically, it was then sealed with concrete, plastic sheeting and/or "crusher run" (one of the finer grades of gravel) before the affected area was repaved. Adjacent ditches were also paved to keep surface water from infiltrating the soils and inducing additional collapse features.

A recent study of 72 sinkhole collapse features along Tennessee highways indicated that 74 percent of the collapses occurred along ditch lines (Moore 1987: 115). Of that 74 percent, 93 percent were found to occur along bare soil or grass-sodded ditches. Because of these conclusions, roadway ditches are now routinely paved in areas of karst terrain.

Although designers prefer to avoid karst areas for new transportation facilities, the highway needs of an expanding population sometimes force construction on undesirable land (Fig. 62). Using karst areas requires proper investigation and design, but even with the best construction techniques there is no guarantee that problems leading to the development of subsidence and induced sinkholes will not occur.

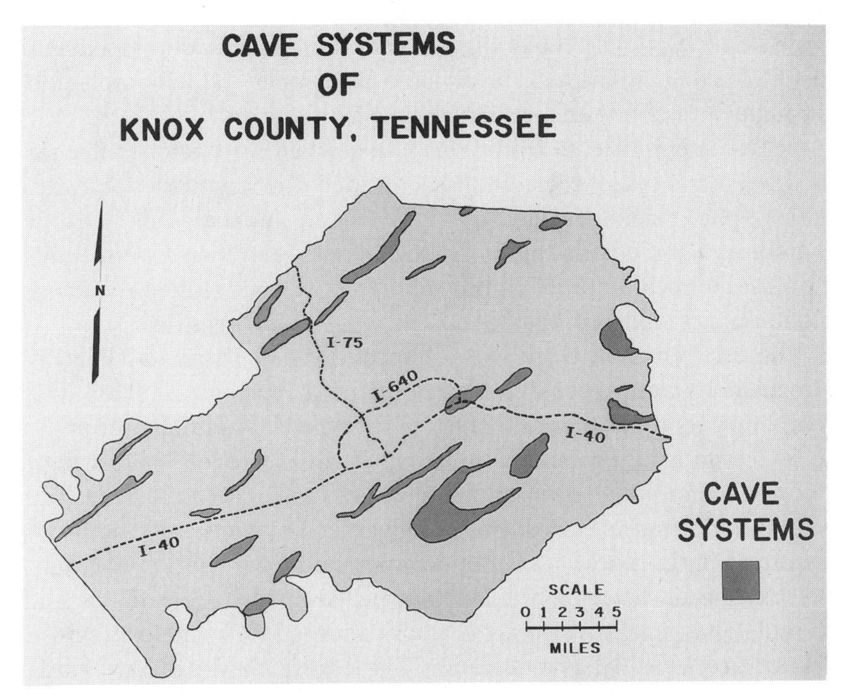

Fig. 62. Cave systems often reflect the locations of karst areas, as illustrated on this map of Knox County. Note the route of I-40 across the county.

I-40 rockfall area on Crab Orchard Mountain. One section of I-40 traverses a mountainous stretch of the Cumberland Plateau known as the Cumberland Mountains (also Crab Orchard Mountains). The rock strata are composed of Pennsylvanian sandstone, shale, and coal beds and Mississippian limestone and shale.

Structurally, these rocks are flat-lying except for a narrow zone of intense folding and faulting (near Mile 331). This folded structure is associated with the Ozone Decollement, a series of splay-type thrust faults (small faults spreading outward from a major fault) emplaced near the end of the Paleozoic era.

The original construction of this two-mile section (Miles 333.4–331.4) through the rugged Cumberland Mountains resulted in numerous vertical roadway cutslopes through horizontal beds of sandstone and shale strata. Because of the difference in hardness between the sandstone (or, in places, limestone) and the shale strata, the layers of harder rock were undermined by the erosion of the weaker shale (a phenomenon called differential erosion or differential weathering). As a result, large blocks of sandstone or limestone come to overhang the lower and more erodible shale slopes. Rockfall problems arise when these blocks, developed as a result of jointing in the rock, break loose along joints and fall or roll onto the highway, creating an immense danger for motorists.

In December 1986, a driver was killed when his tractor-trailer rig skidded into a pile of rock that had cascaded down onto the roadway. A two-phase remedial project was begun immediately: first, a temporary solution, scaling of the rock slopes, and then a permanent cure for the rockfall problem (Fig. 63). The two-mile section requiring attention included both eastbound and westbound lanes.

The first remedial work was a temporary rock slope stabilization effort; more comprehensive and permanent repair came later. The temporary remedial work involved vertical cutslopes ranging from 30 to 50 feet in height, with one precipitous cutslope over 300 feet high (located along the westbound lanes where the fatality occurred). The scaling and trimming effort was accomplished by numerous methods, requiring large backhoes, crane-drawn bulldozer tracks, a crane-hoisted "headache" ball, bulldozers winched up and down the face of the cutslopes, steel-toothed dozer tracks winched back and forth across the cut face, and finally hand scaling. The large quantities of rock debris generated by this unprecedented trimming and scaling effort required special care in hauling and disposal (Fig. 64).

Fig. 63. Rock falls such as this were a traffic hazard along I-40 in Cumberland County (Log Mile 332) until the Tennessee D.O.T. redesigned the cut slopes along this section of roadway.

The final and permanent repair work, begun in early 1987, was completed in December 1988, at a cost of approximately $5.7 million (Moore 1990). The repair flattened the original vertical cutslopes to 1.5:1 and 2:1 slopes (approximately 27° to 34°) and relocated a 1,000-foot segment of the interstate approximately 60 feet away from the 300-foot-high cutslope (Fig. 65). The cost of flattening the high cutslope would have been prohibitive. As a result of the improvements, this section of highway is currently relatively free from rockfall and landslide problems.

Construction of the flattened cutslopes was accomplished by conventional excavation procedures using bulldozers, pans, and, where necessary, blasting. A total of 1,311,767 cubic yards of rock and soil was excavated from the cut intervals and placed in designated waste areas (Moore 1990).

Along this section of I-40 today, you will see mostly grassy cutslopes and notice no signs of the tremendous rockfall problems (Fig. 66). The mitigation of the rockfall hazard is the result of cooperative efforts by the Tennessee Department of Transportation's Geotechnical Section and Construction, Design, and Maintenance divisions.

Fig. 64. Large cranes were used to remove unstable rock along I-40 in Cumberland County in the late 1980s.

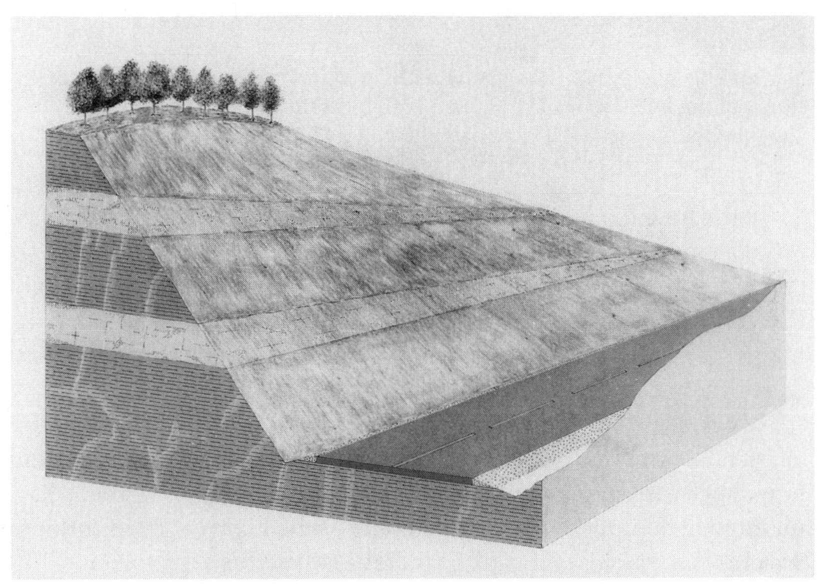

Fig. 65. This set of drawings illustrates the rockfall problem (top) and the remedial cut slope flattening (bottom) along I-40 in Cumberland County (Log Mile 332.5).

Fig. 66. This photograph shows the flatter and smoother reconstructed cut slopes along I-40 (Log Mile 333.0 to 332.0) in Cumberland County on the Cumberland Plateau. TDOT photo by George Hornal.

By the time the public drives on a new highway, many people have put considerable time, engineering skill, and personal effort into the roadway project. Obviously, a new highway cannot appear overnight. Planning, funding, design, construction, and maintenance of our highways all require the input of hundreds of people throughout the Department of Transportation.

Now, as you drive down highways to go to work, or to visit family or friends, or to buy groceries, or to go on vacation, you can reflect on how much careful planning, design, and construction these roads have undergone in order to provide you with a safe highway transportation system. Have a pleasant trip as you take I-40 across Tennessee!

PART 2

Road Log and Side Trips

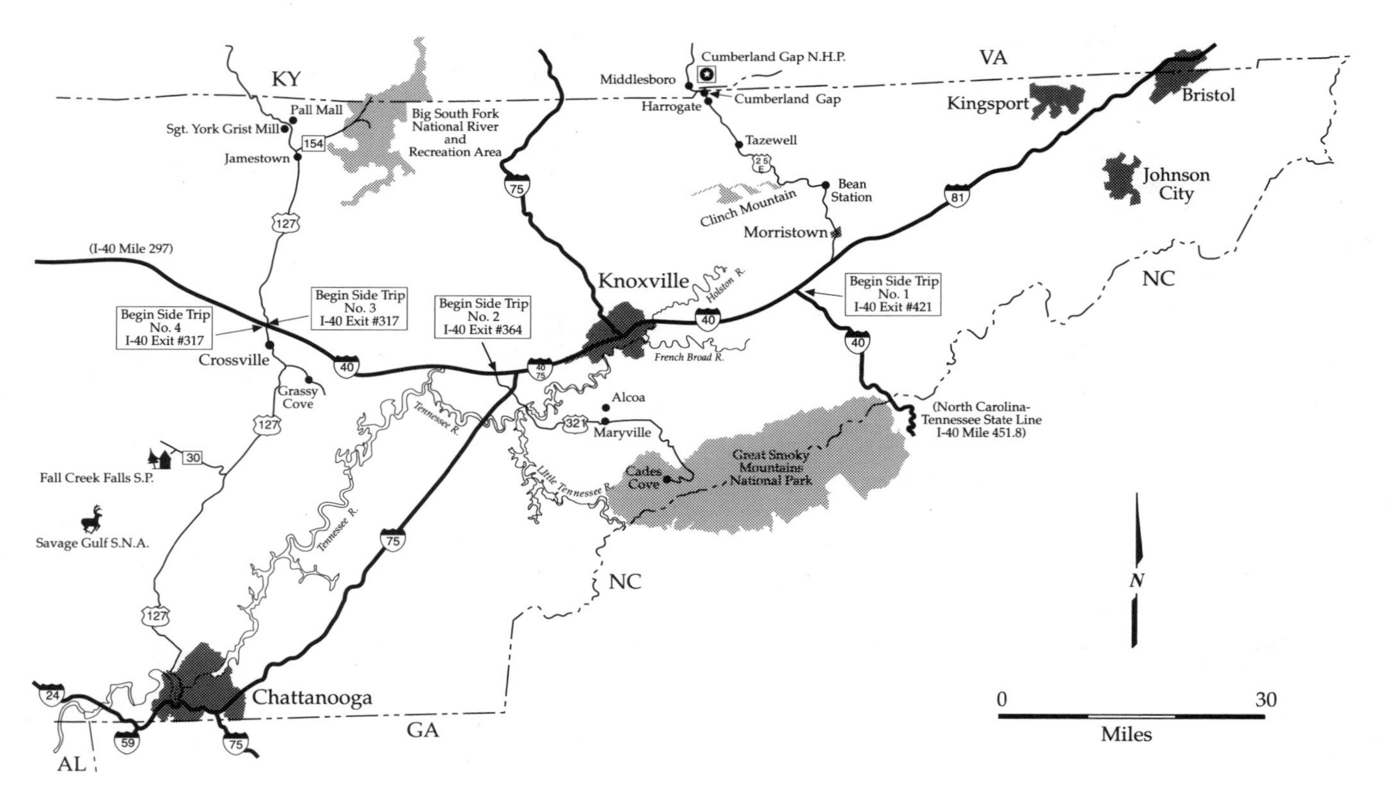

Map 6. East Tennessee, showing interstate highway and locations of Side Trips 1, 2, 3, and 4.

Road Log

East Tennessee

Mile 451.8 (North Carolina–Tennessee state line) to Mile 297.0

Physiographic Provinces

Blue Ridge: Miles 451.8–443.0
Valley and Ridge: Miles 443.0–341.5
Cumberland Plateau: Miles 341.5–297.0

Mileage	Description
451.8	North Carolina–Tennessee state line: Welcome to Tennessee! You have just entered Cocke County (Map 6). For the next 4 miles the area is underlain by metamorphosed Precambrian age sedimentary strata (Ocoee Supergroup; Map 7).

For the next several miles you will be traveling through very mountainous, rugged terrain that is prone to both natural and artificially induced rock slides (Fig. 67). Although this stretch of I-40 is constantly patrolled for rockfall hazards and maintenance, please be aware of the potential for rockfall along this section of highway.

The rocks you will see along I-40 for the next 3 miles have been placed here by large thrust faults, which have moved these rocks over 200 miles from the east-southeast.

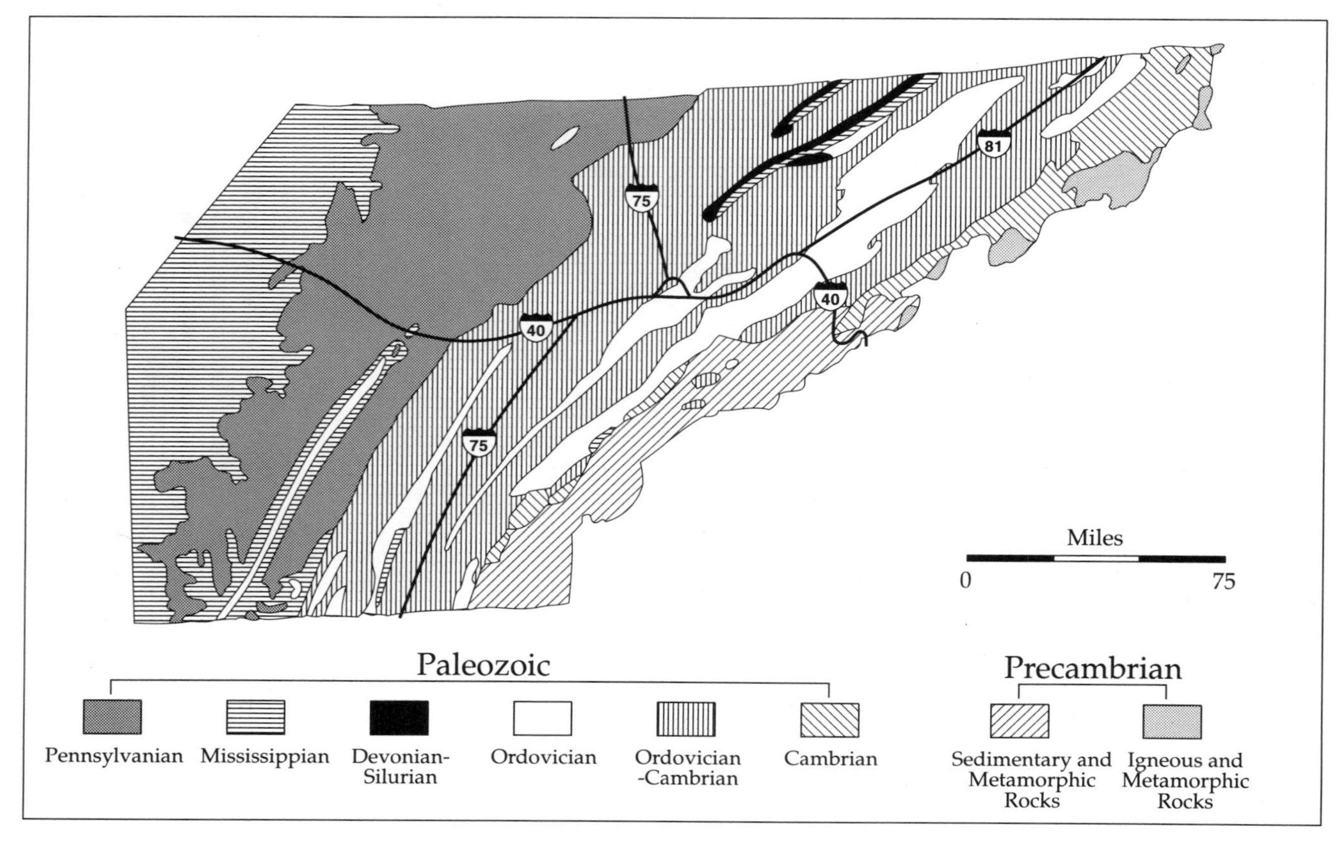

Map 7. Generalized geologic map of East Tennessee. After Geologic Map of Tennessee, Tennessee Division of Geology, 1966.

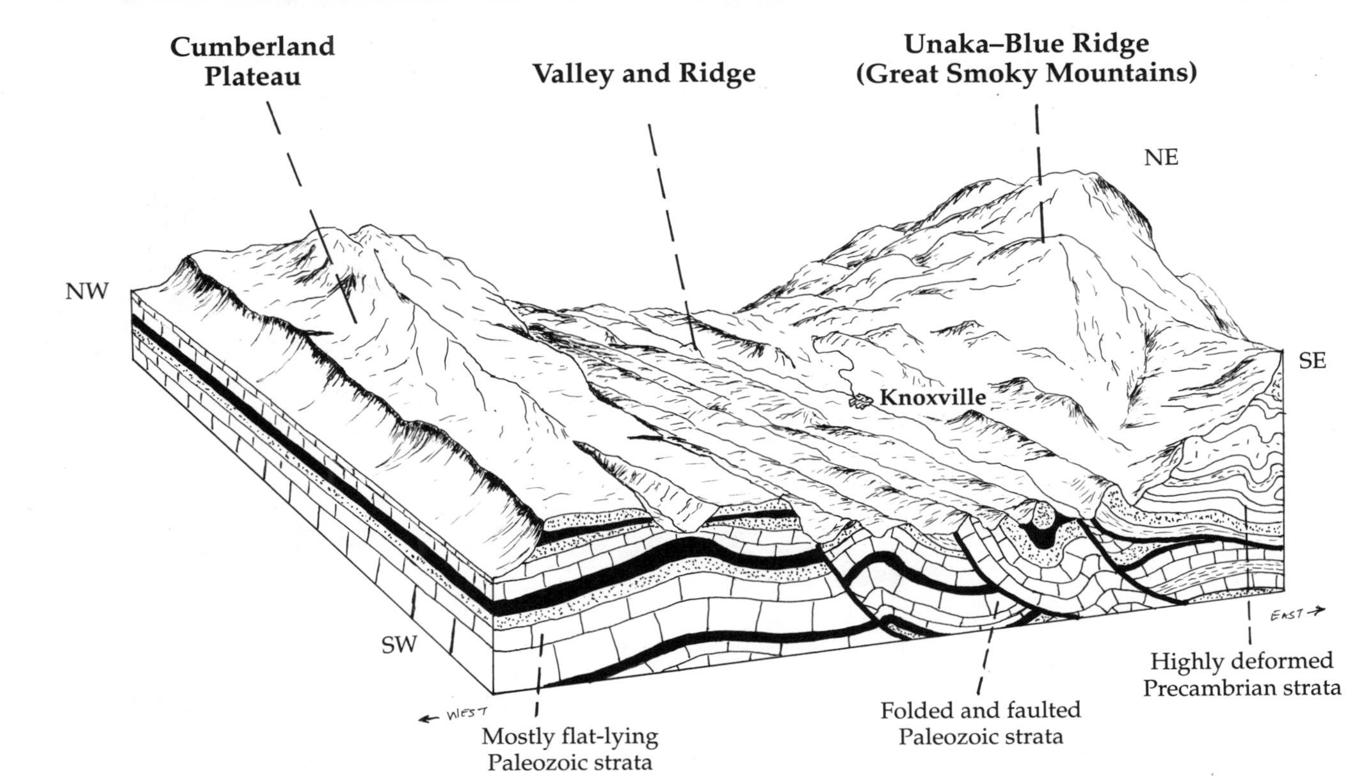

Fig. 67. This schematic block diagram illustrates the complex structural arrangement of the rocks found in the Blue Ridge, Valley and Ridge, and Cumberland Plateau provinces.

451.4 The fence on the westbound lane is to protect against rockfall (Fig. 68). The numerous rock bolts you see protruding along the upper rock slope were used to reinforce the unstable rock slope and tie it to the stable rock mass.

450.6 Exit 451: Waterville interchange. The Appalachian Trail crosses the interstate at this interchange. From the west the trail comes down from Mt. Cammerer in the Great Smoky Mountains National Park, crosses Davenport Gap at the park boundary, and proceeds down the mountain to cross the Pigeon River and I-40 here.

450.2 The Waterville slide scar across the Pigeon River (on your left) resulted from a wedge-failure landslide during the early 1970s.

449.8 The slide area on your right is the site of wedge- and planar-failure landslides that occurred during the 1970s. Remedial work has removed the largest of these slides. The wire mesh hanging on the rock slope is there for protection against rockfall.

449.5 On your right is another slide area that was also the site of wedge and planar failures during the 1970s. Here the remedial work involved removing the total rock mass that was subject to sliding.

Fig. 68. This catchment fence along I-40 (Log Mile 451.4) prevented many tons of falling rock from reaching the highway.

449.2 Yet another slide area is on your right—the site of numerous wedge-failure landslides during the 1970s. Because of rock slides, the interstate had to be closed many times at this spot. The many fractures in the rock that are oriented to the bedding planes at angles other than 90° are known as cleavage fractures. This fractured rock is prone to extensive rockfalls as well as landslides.

448.5 The draped wire mesh on the rock slope to your right is again used to control rockfall (Fig. 69). The catchment fence on the right, at road level, is to protect motorists from falling rocks.

448.2 End of rockfall correction area.

448.0 This is the Hartford slide area, where horizontal drains were used to alleviate a large wedge failure (nearly two million cubic yards of rock is involved). The landslide can be seen along the right ditchline in the westbound lane and along the edge of the river.

447.3 At this point you cross the Dunn Creek thrust fault. The fault trace parallels the valley on the right of I-40 along the base of the high ridge on the right side of the valley.

Fig. 69. Wire mesh, held together with cables, is draped over rock cut slopes to prevent falling rock from hitting the traffic lanes of I-40 (can be seen from Log Mile 450.6 to 448.2).

447.0 Exit 447: Hartford interchange.

446.8 Here you cross the Great Smoky thrust fault. The fault itself is not visible here, but weathering along the fault has produced the long, linear valley in which the interstate is located.

446.4 Tennessee Welcome Center (Fig. 70; off westbound lane). The welcome center, styled after a log cabin, is situated on the toe of an ancient colluvial debris fan that extends up the mountain behind the building. The graveled ditch, located upslope and slightly in the woods, is an interceptor for surface drainage during heavy rains. A note on local culture: Five moonshine stills, representing an era of illegal alcohol production, were found at this location during the construction of the center.

445.1 Pigeon River crossing (sharp curve).

444.3–444.2 Another slide area is on your left. Notice the bedding planes of rock (Cambrian age Chilhowee Group) dipping into the roadway. Because blocks of rock slide on the smooth surface of dipping rock, there is danger of falling rocks. A catchment fence, located along the eastbound lane, is designed to protect motorists from rockfall.

Fig. 70. The Tennessee Welcome Center (Log Mile 446.4) is located on a colluvial debris fan that was part of a large landslide many thousands of years ago.

443.8 Along the left side of the interstate, notice the accumulation of gray to light brown rectangular blocks of sandstone: this is a talus accumulation. Across the Pigeon River canyon, to the right, you can see the dipping layers of the quartzite strata of the Chilhowee Group (the dip is to the northwest at 25°).

443.6 On your left is a colluvial slide area. (There is a dramatic view of Mt. Cammerer from the eastbound lane.) The reddish-orange soil is the residuum of a weathered limestone. The landslide stops just before reaching the interstate.

443.2 Exit 443: Foothills Parkway interchange. (Take this exit if you would like a short, scenic excursion into the mountains. This section of the Foothills Parkway traverses Green Mountain and connects with the Cosby area. At the top of Green Mountain you get a breathtaking overlook into the Cosby Valley and the Great Smoky Mountains that is well worth the extra ten minutes or so that this brief detour from I-40 will cost you.)

443.0 Here you cross the Pigeon River again. This river has been the focus of environmental groups who say the river and its life are being destroyed by the activity of a paper plant in western North Carolina.

442.5 In a cut on your right is an exposure of Cambrian/Ordovician age rock (Knox Group). Here you enter an area of Valley and Ridge topography and Paleozoic age rocks.

440.2 Exit 440: Wilton Springs.

440.0 View of Blue Ridge province (to the rear or from eastbound lane).

439.5 Pigeon River.

438.4 The Pigeon River floodplain is to your right; farmers use its fertile soil for growing produce. A sinkhole area is to your left, above the interstate.

436.5 The light gray rock exposed in the cutslope on your right is Cambrian and Ordovician limestone of the Knox Group.

435.0 Exit 435: S.R. 32 and Newport interchange. Reddish-orange, cherty soil of the Knox Formation underlies this area.

430.0 This area is underlain by Ordovician age strata of Sevier Shale.

429.9 Leave Cocke County; enter Jefferson County.

426.0 Along both sides of I-40 you can see strata of Sevier Shale exposed in the cutslopes: light brown to grayish-brown, thin soil and chippy shale.

425.8 Rest area off westbound lane (with restrooms).

425.0 Douglas Lake (I-40 bridge). More Cambrian/Ordovician age rocks (Knox Group) are exposed in the bluff on the right, along the lake. The medium-bedded strata are limestone and dolostone.

424.0 Exit 424: White Pine/S.R. 113 interchange. Notice the Knox Group outcrops near the I-40 bridge along the right of the road.

421.0 Exit 421: I-40/I-81 interchange. (For Side Trip 1 to Cumberland Gap National Historic Park via Clinch Mountain, which begins here, take the exit to I-81 North.)

419.5 Rest area off eastbound lane.

418.0 The small pond to the left of I-40 is in a sinkhole (rock strata of the Knox Group underlie the sinkhole but cannot be seen from the interstate).

417.9 Exit 417: S.R. 92, Jefferson City, Dandridge.

416.0 Beginning at this point, I-40 becomes three lanes to accommodate the increased traffic approaching metropolitan Knoxville.

415.3 Exit 415: U.S. 25W/70.

414.0 Dipping into the hill on your left are rock strata of the Cambrian age Conasauga Group—dark gray shale overlying light gray limestone.

413.0 Notice the ripple marks in the tilted Cambrian age limestone (Conasauga Group) along the right ditch off the road. (This rock, which dips toward the road at about 35°, is visible only from the westbound lane.)

412.0 Exit 412: Wilton Springs Road. The limestone alongside the exit ramps is Cambrian age (Conasauga Group). The cutslopes along the interstate are in residual clay soil. There is a long valley parallel to the surface exposures of the rock—shale in the valley itself and limestone along its flanks. The ridge south of I-40 (to the left of the road), is underlain by Cambrian age limestone.

409.8 Leave Jefferson County; enter Sevier County.

407.5 Exit 407: S.R. 66, Sevierville, Great Smoky Mountains National Park.

406.3–402.5 Here begins an area underlain by Ordovician age strata of Sevier Shale. This light brown, chippy rock often contains fossils. The cutslopes are excavated on ratios from 1.5:1 to 2:1 (34° to 27°); because of the weathered shale, cutslopes steeper than 1.5:1 would be unstable.

405.0 Leave Sevier County; enter Knox County.

402.0 Ordovician age limestone (Lenoir Formation) is exposed in outcrops near the Midway Road exit (along the left side of the interstate, west of the interchange).

401.5 The reddish-orange residual soil you see is derived from weathering of Knox Group strata.

400.0 The large ridge to your left and straight ahead is underlain by Cambrian/Ordovician age Knox Group strata; there is deep residual soil here.

398.3 Ordovician age limestone and sandstone strata (Holston and Chapman Ridge formations, respectively) are exposed in the cut along the left side of the interstate.

398.8–396.1 This area is underlain by strata of Ordovician age Ottosee Shale.

396.2 Here you can easily get a look at the Ottosee Shale. Look for a small wedge-failure scar in the shale bank by the westbound lane on the right (north) side of I-40. The Ottosee Shale is the light brown to grayish-brown rock.

394.9 Here you cross the Holston River, which flows from upper East Tennessee and joins the French Broad River to form the Tennessee River a few miles downstream from I-40.

394.0 Exit 394: U.S. 11E interchange.

394.0–393.0 Ordovician age shale (Ottosee Formation) is exposed in cutslopes along the right of the interstate. The shale is weathered to a light brownish-gray shaly soil.

392.7 I-640/I-40 interchange; you continue on I-40.

392.5 Ordovician age limestone (Holston Formation) is exposed in the cutslope on your left. Notice the pinkish color of the rock (the famous "Tennessee Marble") and the numerous solution cavities; in the local area many caves are developed in this kind of rock. The rock formation gets its name from the locality: the first description of this rock by geologists was of a bluff on the Holston River about a mile upstream from the point where I-40 crosses the river (Mile 394.9).

392.0 Exit 393: U.S. 11W, Rutledge Pike, Knoxville Zoo. The next 2 miles along I-40 (to both your left and your right) is a karst area underlain by the Holston Formation, a limestone.

390.0 Exit 390: Cherry Street. Notice the tilted, grayish-brown strata of the Ordovician age shale (Ottosee Formation) exposed on your left and right. These rock layers were once horizontal.

388.8 U.S. 441 exit to Knoxville downtown loop (James White Parkway).

388.0 Downtown Knoxville is underlain by Cambrian/Ordovician age limestone and dolostone strata of the Knox Group (the rock strata are not visible from the interstate).

386.4 Exit 386: U.S. 129, airport (McGhee-Tyson), Tennessee Department of Transportation (TDOT) Region I office.

384.7 At the I-40/I-640 interchange, you continue west on I-40.

384.5 Ordovician age limestone (Holston Formation), sandstone (Chapman Ridge Formation), and shale (Ottosee Formation) underlie this area but are not generally visible from the interstate.

379.0 Here you begin to cross a large karst basin. An area of 15 square miles drains into the Ten Mile Creek cave system, approximately 2 miles south of I-40; commercial development of the area has increased the runoff into the Ten Mile Creek drainage.

378.6 Cross Ten Mile Creek (no marker). The orange-red soil you see is derived from weathering of Knox Group rock strata (limestone and dolostone).

376.0 Interchange with Pellissippi Parkway. (The Pellissippi Parkway, which meets S.R. 62 and crosses the Clinch River about 7 miles to the north, leads to the Manhattan Project city of Oak Ridge. In Oak Ridge is the American Museum of Science and Energy. About 7 miles west of Oak Ridge on Bethel Valley Road is the Oak Ridge National Laboratory and the Graphite Reactor, a National Historic Landmark. Bethel Valley Road joins S.R. 95 to return to I-40 at Exit 364.)

374.9 Exit 374: Lovell Road, S.R. 131.

371.1 The large ridge you are descending (and which continues to your left) is underlain by Cambrian/Ordovician age limestone and dolostone strata (Knox Formation). There is deep residual soil at the top of the ridge (about Mile 371.0).

368.8 Leave Knox County; enter Loudon County.

368.1 Interchange with I-75/I-40; continue on I-40.

367.5 On your left and your right, in the cut through the ridge, notice the tilted shale and siltstone strata.

367.0–365.2 This stretch of highway, underlain by Cambrian/Ordovician age strata (Knox Group), has experienced numerous sinkholes (Figs. 71 and 72). The Clinch River is off to your right, out of view; surface drainage is controlled by karst activity.

364.1 Exit 364: U.S. 321 and S.R. 95. (Side Trip 2 to Cades Cove in the Great Smoky Mountains National Park begins here.) One mile north of the exit is the Tennessee Valley Authority (TVA) Melton Hill Dam on the Clinch River.

363.6 Leave Loudon County; enter Roane County.

362.7 Rest area off westbound lane (no restrooms).

362.0–359.0 I-40 turns and follows the valley for several miles. This valley contains many caves, of which Eblen Cave is the most popular with local cavers because of its large entrance (over 100 feet wide).

Fig. 71. This sinkhole collapse occurred near Log Mile 367.0 on the westbound lane of I-40 and is characteristic of sinkhole problems where roads are built over karst landscape. Photo courtesy of Jim Aycock.

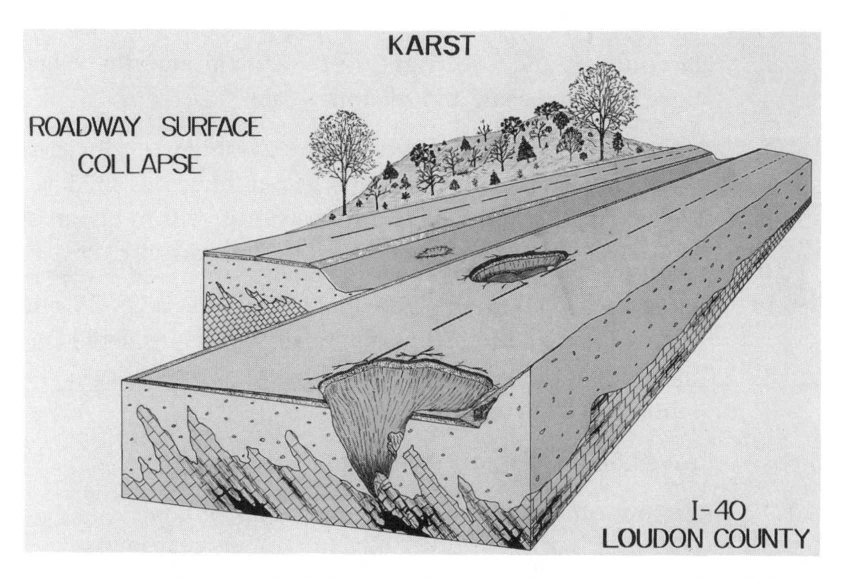

Fig. 72. This schematic block diagram illustrates the occurrence of sinkhole-collapse features along highways, especially I-40 in Loudon County (Log Mile 367.0).

362.0 Rest area off eastbound lane (no restrooms).

360.9 Exit 360: Buttermilk Road.

357.0 The large ridge you are now crossing is underlain by Cambrian/Ordovician limestone and dolostone strata (Knox Group).

353.3 Here you cross another large ridge underlain by strata of the Knox Group.

352.7 Exit 356: S.R. 58, Kingston.

352.0 Strata of the Knox Group underlie this ridge, but only residual clay soil is visible at the surface.

351.7 Here you cross the Clinch River, which flows from southwest Virginia into Tennessee and encounters TVA dams at Norris and Melton Hill before joining the Tennessee River a few miles downstream (to your left).

351.4 On your right is TVA's Kingston Steam Plant (fossil fuel). Here begins a section where the interstate crosses several ridges and valleys.

350.0–349.0 This area is underlain by folded and faulted Cambrian/Ordovician strata; the ridge is underlain by limestone and dolostone strata of the Knox Group.

347.9 Exit 347: U.S. 27, Rockwood and Harriman.

346.5 Now you leave the Valley and Ridge province and enter the Cumberland Plateau. Here, at the bottom of the Cumberland Escarpment, begins the Rockwood Mountain landslide area.

346.0 You begin ascending Rockwood Mountain.

345.4 To your right (westbound lane only) is a very large, active landslide that originated during the construction of I-40, about 1970. Landslide movement is active near the top of the ridge to your right (Walden Ridge). The area next to the interstate is stable.

345.0 Mississippian age strata underlie this section of I-40. (If you are traveling in the eastbound lane, you can see Fort Payne chert, the reddish residual soil in the cut on your right.)

343.3 Gabion baskets are in the median on your left and high up on the slope to the right of the roadway (westbound lane only). These rock-filled wire baskets are holding back unstable colluvial deposits that slid during construction of the interstate.

341.5 End Rockwood Mountain landslide area. You are now at the top of the Cumberland Escarpment and entering the Cumberland Plateau. (If you are traveling in the eastbound lane, you get a breathtaking view of the Valley and Ridge province as I-40 leaves the plateau and begins descending the escarpment. Unfortunately, there are no pull-offs, so you must not stop.)

341.0 The outcrops of sandstone and shale along the left side of the road are Pennsylvanian age rock. (If you are traveling in the eastbound lane, you can see exposures gray shale, sandstone, and black coal that have been gently folded into a downward warp, forming a small syncline.)

340.5 Leave Roane County; enter Cumberland County. Here you also leave the Eastern Time Zone and enter the Central Time Zone.

339.0 Cross Creek.

338.9 Exit 338: Westel Road.

335.8 Parking area off eastbound lane (no restrooms).

334.7 On your left there is a spectacularly high railroad trestle (Norfolk Southern Railway line).

334.0 In the sandstone cut along the right side of the road (westbound lane only), notice the pre-split lines on the rock face. Holes are drilled about 24 to 36 inches apart in a straight line and blasted before the main detonation so as to crack the resulting rock face evenly.

333.4–331.4 Another rockfall landslide repair area begins here. This section of road has experienced numerous rockfalls and small landslides (Fig. 73).

333.2 In the cut to the left and the right of the road, notice the cross-bedded sandstone. The cross-beds (the thin, angular lines that are slightly curved and intersect with other sets of angular lines) are the results of rapidly deposited and advancing deposits of sand (mostly marine). The other, vertical lines you can see in the rock slope are again pre-split lines from precision drilling and blasting to control rockfall.

332.0 In 1986 the large cutslope to the right of the interstate precipitated several rockfalls that covered the roadway and resulted in a fatality.

331.4 End of the highway rock slope repair area.

330.7 Along the right side of the road you can see gray strata of Mississippian limestone that underlie the Crab Orchard cove area.

329.6 Exit 329: Crab Orchard, U.S. 70. The underground quarry to the north extracts Mississippian age limestone. Because of its high concentration of calcium carbonate (usually more than 90 percent $CaCO_3$), the stone is crushed and processed for use in cement and agricultural lime products.

326.7 Rest area off westbound lane (with restrooms).

326.0 You have a dramatic view of Crab Orchard Mountain if you are in the eastbound lane (from the westbound lane you have to turn around and quickly look back to the east to see the mountain).

Fig. 73. Rock falls were numerous along this stretch of I-40 (Log Mile 333.0 to 331.0) until the Tennessee Department of Transportation completed a remedial project in 1989.

324.3 Rest area off eastbound lane (with restrooms).

323.8 The reddish-orange sandstone outcrops along both sides of the road are Pennsylvanian strata. Called Crab Orchard Sandstone, it is quarried locally as a building stone.

322.2 Exit 322: S.R. 101.

321.5 To the west you can see the flat, tabletop effect of the plateau area. The outcrops of black and gray shale exposed in the road cut to the left and the right at the top of the hill are of Pennsylvanian age.

320.0 Exit 320: Genesis Road and S.R. 298.

318.1 The sandstone outcrops located along the left side of the highway (visible from both lanes) are of Pennsylvanian age.

318.0 Cross the Obed River. The Obed, designated a National Scenic River, flows from the south (left) to the north (right).

317.8 Exit 317: Crossville, U.S. 127. (You take this exit for both Side Trip 3 to Twin Arches in the Big South Fork National River and Recreation Area and Side Trip 4 to Fall Creek Falls State Park, as well as for Pickett State Forest.)

313.0 Here begins an area of large, deeply eroded, sloping hills with intervening flat stretches—a rolling landscape.

311.0 Exit 311: Plateau Road.

306.7 Rest area (no restrooms). There are few rock outcrops along this stretch of I-40, which travels through a heavily forested area.

304.6 Leave Cumberland County; enter Putnam County. The areas of shallow-rolling to level land along this section of the interstate reflect the flat-lying structure of the underlying strata.

304.0 To the right of the highway is a sandstone exposure of Pennsylvanian age.

303.5 The black shale exposed along the right side of the road is Pennsylvanian in age. Long ago, when this rock was deposited, the area was probably a lagoon.

301.4 Exit 301: S.R. 84. To your right (north) you can sometimes get a fine view of the adjacent valley and the Cumberland Escarpment. (During the spring and summer, the scenic view is usually obscured by vegetation, but late fall and winter provide good opportunities.) Please note that this is not a designated parking area, and you must not stop.

300.4 The brownish-gray sandstone outcroppings along the right side of the westbound lane are Pennsylvanian in age.

298.8 In the outcroppings along the right side of the interstate, you can see a good example of a channel deposit in Pennsylvanian age strata (Fig. 74). Notice the curved boundary between the black shale and tan sandstone. The sandstone represents a stream channel that cut through a mud flat in an ancient lagoon. The curved lines you see within the sandstone are again cross-beds.

Fig. 74. Near Milepost 298.8, I-40 passes along exposures of Pennsylvanian-age rock strata. Here the light-colored sandstone represents a channel of an ancient stream that cut through possible lagoon sediments (dark gray to black).

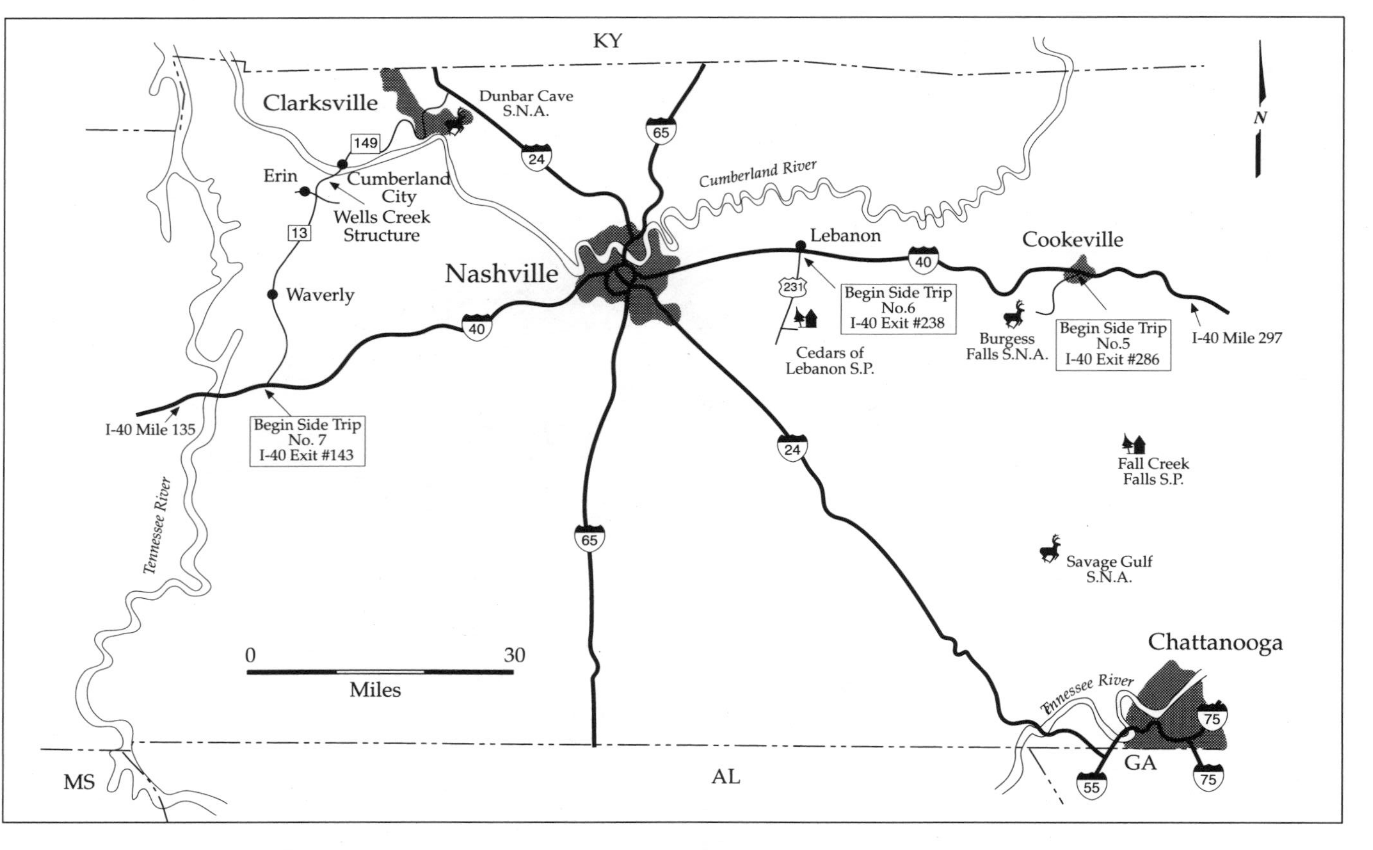

Map 8. Middle Tennessee, showing interstate highway and locations of Side Trips 5, 6, and 7.

Middle Tennessee

Mile 297.0 to Mile 135.0

Physiographic Provinces

Eastern Highland Rim: Miles 297.0–272.5
Central Basin: Miles 272.5–186.0
Western Highland Rim: Miles 186.0–135.0

297.0 Here you are at the top edge of the Cumberland Escarpment, and the descent from the Cumberland Plateau onto the Highland Rim is rather steep (Map 8). The top edge of the escarpment is underlain by Pennsylvanian sandstone, shale, and coal (Map 9). As you continue down the escarpment, the rock strata become older—Mississippian in age.

296.0 Along both the right and the left sides of the highway, you can see good exposures of Mississippian age strata (Pennington Formation): flat-lying green and maroon shale and grayish limestone. You are now beginning your descent off the Cumberland Escarpment onto the Highland Rim.

295.2 As you continue your descent, notice the massive, light gray Mississippian limestone exposed in the cuts along both sides of the road.

295.0 The limestone and green shale you see along the left side of the highway are more exposures of the Pennington Formation.

294.0 The pinnacles of limestone exposed in the cutslope along the right side of the road were caused by weathering.

292.1 You are now on the Highland Rim (Fig. 75). On your left, across the interstate lanes, are large limestone outcrops (Mississippian age); notice the lower shale layer and the upper limestone, with some springs issuing from solution cavities. For the next 20 miles the landscape is underlain by Mississippian limestone strata.

291.5 Cross Falling Water River. This river continues its passage across the Highland Rim and forms Burgess Falls, where it cascades off the Highland Rim escarpment onto the Nashville Basin.

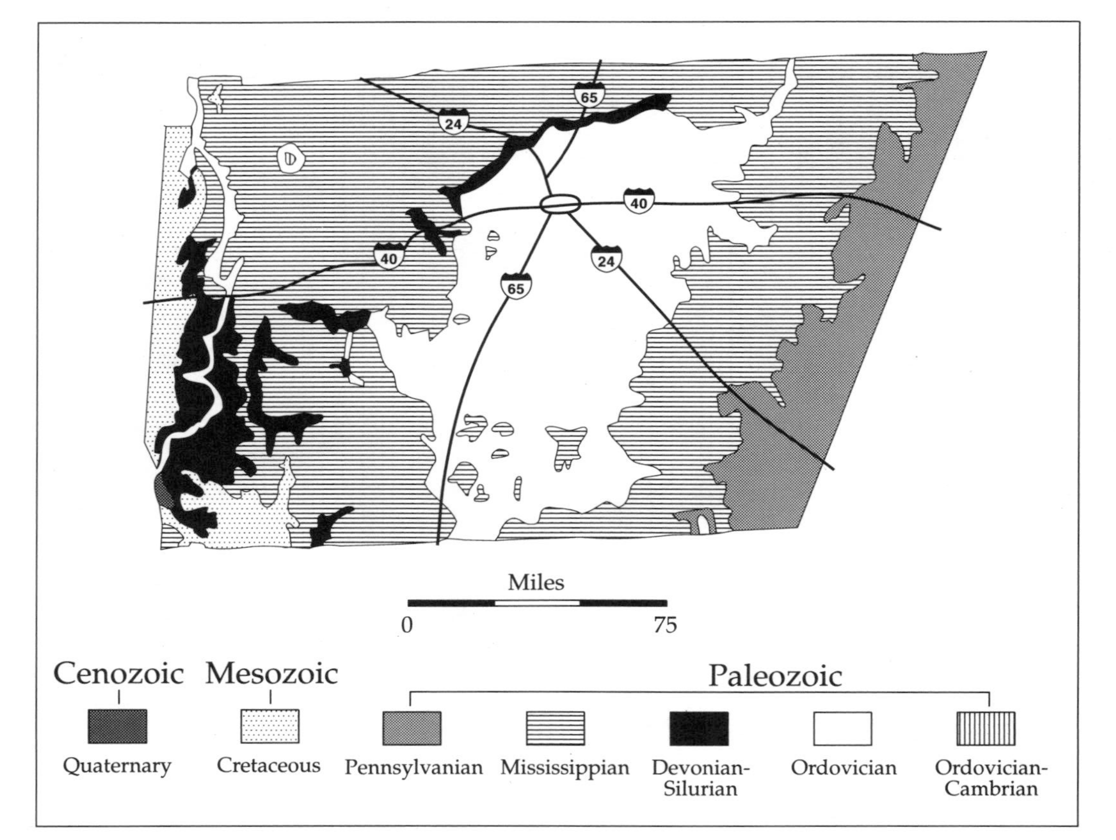

Map 9. Generalized geologic map of Middle Tennessee. After Geologic Map of Tennessee, Tennessee Division of Geology, 1966.

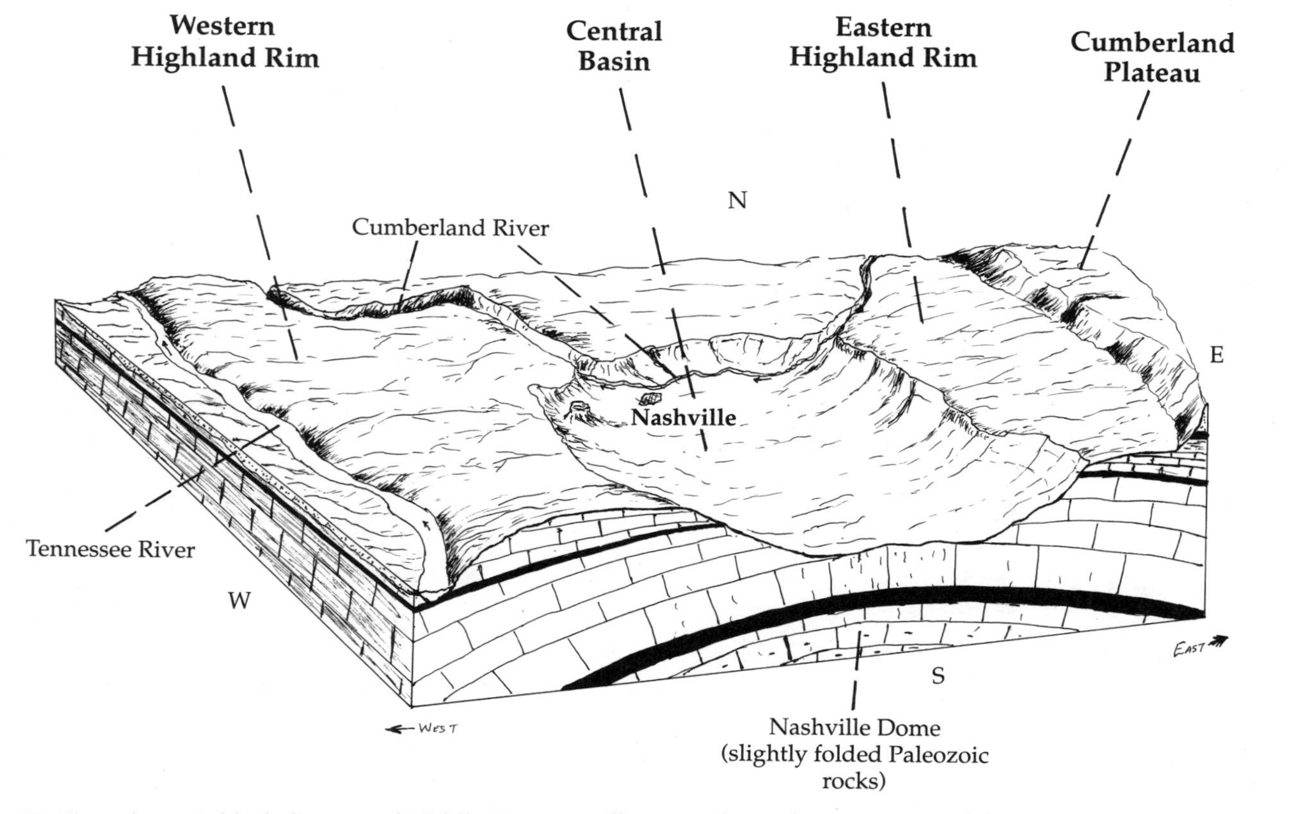

Fig. 75. This schematic block diagram of Middle Tennessee illustrates the geologic structure of the Eastern Highland Rim, Central Basin, and Western Highland Rim provinces.

290.5 Here the Cumberland Escarpment can be seen (from eastbound lane only).

290.4 Exit 290: U.S. 70.

289.0 Notice the quarry face on the north side of the highway, where Mississippian limestone strata are exposed. The numerous sinkholes, which you may notice in the landscape, indicate that the underlying limestone is slowly being dissolved by the acidic groundwater conditions.

288.8 Exit 288: S.R. 111.

287.5 If you are eastbound rather than westbound, you can see Mississippian age limestone outcrops to your left from the eastbound lane. (They are not visible from the westbound lane.)

287.0 Exit 287: Cookeville, S.R. 136; TDOT and Tennessee Highway Patrol offices.

285.4 Exit 286: S.R. 135. (You take this exit for Side Trip 5 to Burgess Falls State Natural Area.)

280.1 Exit 280: S.R. 56.

273.4 Exit 273: Another exit to S.R. 56. The stretch of highway you have driven from Cookeville to the Smithville exit is very scenic, with farms and woods (but no outcrops to observe). This is the Eastern Highland Rim.

272.5 You have reached the top of the Highland Rim escarpment and now begin your descent from the Highland Rim onto the Central Basin, which is underlain by limestone. Here you see rolling hills and deep gorges, but few outcrops; the soil is reddish clay and chert. As you descend off the Highland Rim, the change in landscape is dramatic and very scenic. To the left of the interstate, you can see a beautiful valley flanked by the steep, grassy slopes of the escarpment.

The interstate has a median barrier for about 1.2 miles along this stretch. In order to avoid making more massive cuts through the hills and having to put fills in the valley, the interstate width was reduced by eliminating the usual wide, grassy, divided median. In this way, damage to the scenic environment was reduced.

271.7 At this point there is contact between the Mississippian age Fort Payne Formation (a silicious limestone and bedded chert) and Devonian age shale (Chattanooga Formation; Fig. 76). The Chattanooga Formation, which is black and very thinly laminated, can be easily identified; sometimes the black shale has rusty stains on its exposed surface because of oxidation of the mineral pyrite (an iron sulfide), also known as fool's gold. As you continue down the escarpment, you will notice several large exposures of limestone strata.

Lying at the base of the Fort Payne strata and on top of the Chattanooga is the Maury Shale, a light greenish-gray shale of Mississippian age. The Fort Payne Formation, in beds 1 to 2 feet thick, is a yellowish-gray, silicious limestone.

270.9 To the right of the westbound lane you can see a large exposure of limestone strata. The rock near the top of the cutslope is Devonian age black shale (Chattanooga Shale) and that near the roadway level is Ordovician age limestone.

270.4 Again you can get a look at a large exposure of Ordovician limestone strata to the right of the interstate. The bedding of

Fig. 76. Near Log Mile 272.5, Interstate 40 cuts a path from the Eastern Highland Rim into the Central Basin exposing Mississippian-, Devonian-, and Ordovician-age limestone. The thinly layered black shale is the Devonian-age Chattanooga Formation.

this medium gray rock ranges from thin (a few inches thick) to medium (about a foot thick).

270.0 To the right of the westbound lane is another large exposure of limestone strata (over 100 feet high). These strata have undergone some solution activity from the groundwater. In the rock slope near Mile 269.8, you can see a large "cutter" (a place where soil has filled in between bedrock pinnacles); in this case, the highway excavation has cut across a soil-filled sinkhole (10 to 12 feet across) and the soil has partially eroded out, leaving a dish-shaped scar in the soil behind the rock face.

269.0 Again, you can get a look at a large exposure of thin- to medium-bedded Ordovician limestone strata, over 100 feet high, to the right of the interstate. The hilly terrain here posed a problem for the highway construction. Maintaining the proper gradient of the roadway as it wound down the Highland Rim escarpment required some extensive cutting of the hills, and that resulted in some rather high cutslopes. Accordingly, flat benches were cut into the rock slope here to keep the cutslope stable and to provide catchment areas for rockfall as the fresh exposure weathered. Notice that cedar trees, which prefer very thin, poor soil, have become established on the flat bench areas of the cut.

268.3 Exit 268: Center Hill Dam. This dam (not visible from the interstate) was constructed and is operated by the U.S. Army Corps of Engineers.

268.2 Large exposure of limestone strata.

268.0 Now you begin to traverse the Central Basin, which is underlain primarily by Ordovician age rock. Most of the strata in the Central Basin consist of limestone, with some shale formations.

267.3 Caney Fork River; rest area (with restrooms). If you stop at this rest area, you may wish to take a little time to stroll around the grounds. The Caney Fork River passes within a few hundred feet of the rest area. The impressive bluff you see along the river is composed of Ordovician limestone. Notice that the rock layers are horizontal and, unlike those found in East Tennessee, have not been tilted. *Please note:* The river waters are subject to quick and sudden changes in flow and depth due to periodic discharges from Center Hill Dam, so be careful!

267.2 Leave Putnam County; enter Smith County.

266.8 Cross Caney Fork River. As you continue to drive west on I-40, you will notice that the highway crosses the Caney Fork River many times. The river winds (meanders) along the base of the outlying hills of the Highland Rim, making several large, sharp bends. The highway, with its straight course, cuts across these bends, giving you opportunities to see the course of this scenic river. You may also have a chance to spot our national bird, the bald eagle, along the Caney Fork (in recent years, the birds have been reintroduced in several areas of the state, including Center Hill).

266.4 Cross Caney Fork River.

265.6 Cross Caney Fork River.

264.5 Along the left side of the road (westbound lane only), you can see large, medium gray outcrops of Ordovician age limestone strata of the Nashville Group, some of them massive.

263.9 Cross Caney Fork River.

263.1 Cross Caney Fork River.

261.0 Along the westbound lane the topography is becoming less hilly and more rolling (in the eastbound lane, however, you begin to notice higher hills and knobs).

259.2 The light gray outcrops are Ordovician age limestone, the bedrock in this area. This limestone has a shaly appearance because it is argillaceous (that is, it has a high clay content).

258.6 Exit 258: S.R. 53, Cordell Hull Dam. This U.S. Corps of Engineers dam is named for a famous Tennessean: a U.S. congressman and later secretary of state, who won the Nobel Peace Prize in 1945 for his role in establishing the United Nations.

257.5 Along the right side of the westbound lane are Ordovician age strata (Nashville Group) consisting of medium-bedded gray limestone underlain by gray shale or shaly limestone. Here you can observe a phenomenon called differential weathering: because the shale is softer than the limestone, it weathers faster, and there is a pronounced boundary between the two types of rock.

256.1 The medium to light gray limestone outcrops on both the left and right of the highway exhibit what is known as joint-controlled solution activity. The vertical cracks in the limestone, the joints, have been enlarged because of the solution activity of the groundwater.

254.4 Exit 254: S.R. 141.

252.6 Rest area (no restrooms).

252.0 Along the right side of the highway, you can observe more shaly Ordovician limestone outcrops of the Nashville Group.

251.8 The large limestone outcrops are strata of Ordovician age belonging to the Lebanon and Carters formations (Stones River Group) and the Hermitage Formation (Nashville Group). You will see more occurrences of this grayish-brown, shaly limestone at various points along the interstate for the next 10 miles or so (noted individually below).

250.5 To the right are more Ordovician limestone outcroppings.

250.1 Leave Smith County; enter Wilson County.

249.5 The limestone outcroppings to the right of the westbound lane are again Ordovician in age. These shaly (argillaceous), thin-bedded strata belong to the Stones River Group. Notice that the strata are slightly tilted, dipping west-northwest a few degrees from horizontal.

248.5 The Ordovician limestone outcroppings to both the left and the right of the interstate are strata of the Stones River Group.

246.0 Exposed to the left and right of the interstate are thin-bedded, shaly strata of the Nashville Group (more Ordovician limestone).

245.7 Once again, the bedding of the Ordovician limestone outcroppings you see on the left and the right of the highway is thin and horizontal.

245.0 Exit 245: Linwood Road.

243.5 The Ordovician limestone outcroppings along the right side of the interstate are Lebanon Limestone of the Stones River Group.

241.6 You can observe more horizontal-bedded strata of the Stones River Group to the left and the right of the interstate.

239.8 Exit 239: U.S. 70. Here you see low, rolling hills that are underlain by Ordovician limestone; soil is thin and acid.

238.0 Exit 238: Lebanon and Cedars of Lebanon State Park. Although the area is underlain by Ordovician limestone, this stretch of highway has few outcrops; those are mostly shaly Lebanon Limestone of the Stones River Group. (Side Trip 6 to Cedars of Lebanon State Park begins at this exit.)

231.9 Exit 232: Gallatin, S.R. 109. For the next 8 miles, the upland areas along both sides of the interstate are capped by the shaly limestone of the Hermitage Formation, while the interspersed valleys are underlain by limestone of the Carters Formation.

228.0 Parking area off westbound lane (no restrooms).

226.0 Exit 226: S.R. 171, Mt. Juliet Road, Long Hunter State Park. (Day-use recreational activities are provided at Long Hunter State Recreational Park, which is located along the shore of Percy Priest Reservoir, just east of Nashville.)

222.7 Leave Wilson County; enter Davidson County.

221.0 Exit 221: S.R. 45. In 1969, approximately 6 miles to the north of this exit, the Du Pont Company at Old Hickory, Tennessee, drilled a test well that encountered Precambrian basement rocks at 5,460 feet below the surface (R. L. Wilson 1981: 45).

220.5 The Ordovician limestone outcrops along the left side of the interstate are of the Hermitage Formation. This thin-bedded, shaly limestone was so named because the first description of it was of rock observed near Hermitage, Tennessee (a community named for the nearby home of President Andrew Jackson).

219.5 Percy Priest Dam and Stones River. The dam was constructed and is operated by the U.S. Army Corps of Engineers. Below the dam you may be able to see outcrops of Carters Limestone in the riverbed during times of low water.

216.8 Exit 216: Airport and S.R. 255. The rocks you see along both sides of the interstate are outcrops of the Hermitage Formation.

215.0 Exit 215: S.R. 155, Briley Parkway, Opryland.

213.5 Nashville area. At the I-40 Interchange with I-24, you can again see the Hermitage Formation exposed in roadway rock cuts.

213.2 As you continue west on I-40, you can see Hermitage limestone exposed in the ramps for the interchange with I-24. At this location the Hermitage Formation consists of ribboned, silty, laminated limestone strata.

213.0 I-24/I-40 split; proceed on I-40 West. (An alternate route is I-240 West to bypass downtown Nashville. If you choose I-240, which adds about 8 miles, you will rejoin I-40 at Mile 206.0, below.)

212.0 Here you can see the gray, horizontal strata of Ordovician limestone (Nashville Group) exposed to your left and right.

211.2 Continue on I-40, which turns to your left. (The I-65 North/I-24 West exit is to your right.)

210.0 I-65 South/I-40 interchange. The hill at this point is underlain by Ordovician age strata. The rocks exposed in the interstate cuts (left and right) are thick-bedded limestone of the Bigby-Cannon Formation (Nashville Group).

208.5 Notice the Nashville skyline, with the State Capitol on your right (206 feet 7 inches in height to the top of its tower). Designed by William Strickland, and built between 1845 and 1854, the Tennessee State Capitol is a National Historical Landmark. Because the original stone (quarried only about a mile from the location of the building) had deteriorated, the State Capitol was renovated in 1956, using about 90,000 cubic feet of Indiana limestone, chosen to approximate the character of the earlier facing.

207.9 I-265/I-40 Interchange. All the strata exposed in the cutslope to the left of the interstate are of Ordovician age (Nashville Group).

206.5 Ordovician strata of the Nashville Group are again exposed in outcrops at Exit 207.

206.0 At the I-240/I-40 interchange you can see Ordovician limestone exposed in the big cut on your left. These thin-bedded, shaly strata belong to the Nashville Group.

204.1 Exit 204: White Bridge Road; S.R. 155. Strata of the Hermitage Formation are exposed along both the left and the right sides of I-40, just west of the White Bridge Road overpass bridge.

201.5 Where the interstate cuts through this hill, Ordovician strata belonging to the Bigby-Cannon Formation of the Nashville Group are exposed.

201.2 Exit 201: U.S. 70, Charlotte Pike. You can see limestone of the Bigby-Cannon Formation along the ramps down to Charlotte Pike.

199.4 Exit 199: Old Hickory Boulevard, S.R. 251.

199.0 This area is underlain by the gray, horizontally bedded Ordovician limestone of the Nashville Group.

197.6 The flat-lying, shaly limestone in the exposures on your left and right is of Ordovician age. The strata belong to the Catheys Formation and the Nashville Group.

196.4 Exit 196: U.S. 70S.

194.7 Cross Harpeth River. Designated a Scenic River, the Harpeth is very popular with canoeists.

194.0 Here you see the Harpeth River floodplain, an area representing extensive flooding and the deposition of very fertile silt and sandy soil.

193.3 Notice the rock exposures on your left and right. The gray strata in the upper portion of the cutslope are Mississippian age limestone. The black layer, 10 to 15 feet thick and quite fissile, is Devonian age shale (Chattanooga Formation). The greenish-gray rock in the lower part of the cut is Silurian age shale (Waldron Formation), which is very fossiliferous.

192.9 To the right and left of the interstate you can see another exposure of Silurian age shale (Waldron Formation).

192.8 Exit 192: McCrory Lane, Pegram.

191.6 Leave Davidson County; enter Cheatham County.

191.0 The black shale that you see along the right of the road in the large cutslope is the Devonian age Chattanooga Formation.

190.5 On your right is another exposure of limestone and shale. The black shale is again the Chattanooga Formation; the gray limestone above it is the Mississippian age Fort Payne Formation.

190.1 Once again you cross Harpeth River, which meanders along the base of several outlying hills of the Western Highland Rim.

188.6 As you cross the Harpeth, notice the massive bluff on your left. Most of the bluff that you see is formed from the Mississippian age Fort Payne Formation (a silicious limestone) and Warsaw Limestone.

188.2 Exit 188: Kingston Springs. Quartz crystals can be found in small, open cavities (vugs) in the strata exposed along the entrance ramp to I-40 West (Fig. 77).

187.0 Here you begin the ascent onto the Western Highland Rim. In the numerous exposures of the Mississippian age Fort Payne Formation, to both the left and the right of the interstate, you can see the horizontal structure of the bedrock. Near the top of the hill straight ahead are exposures of another Mississippian limestone, the Warsaw Formation, which is less silicious than the underlying Fort Payne Formation.

186.0 Now you begin to traverse the Western Highland Rim area.

185.8 The Kingston Springs lookout tower, on your right, is located on a hill that is capped by the Warsaw Formation.

184.4 Leave Cheatham County; enter Williamson County.

181.8 Exit 182: S.R. 96, Montgomery Bell State Park. The top of this hill is underlain by Mississippian age strata (Warsaw Formation). You can see two small, weathered exposures of the Warsaw Formation along I-40, where S.R. 96 crosses the interstate.

181.2 Leave Williamson County; enter Dickson County.

179.5 If you are eastbound, to your left you can see massive outcrops of the Mississippian age Fort Payne Formation.

Fig. 77. The Mississippian-age Fort Payne Formation contains numerous deposits of silica commonly in the form of quartz-filled vugs. These are easily seen in the ramp to I-40 West at Exit 188.

Notice the horizontal layers. (These outcroppings are not visible from the westbound lane.)

177.0 In the cutslope along your right (westbound lane only), you can see more strata of the Fort Payne Formation.

172.7 Exit 172: Montgomery Bell State Park, S.R. 46.

170.3 Rest area (with restrooms).

166.4 As you cross the Piney River, you can see exposures of the Fort Payne Formation in the riverbed.

163.6 Exit 163: S.R. 48.

163.3 Leave Dickson County; enter Hickman County.

162.5 Here you cross the Garner Creek drainage area, where erosion has cut a deep incision into the hard, eroded clay soil that results from the weathering of the Fort Payne Formation limestone. Notice that in the road cuts the bedrock is not visible: the limestone is soluble, forming clay soil.

156.0 The cherty, reddish soil you see is residual clay from limestone.

155.0 The limestone cliffs to your right (also visible from the eastbound lane) are Mississippian age strata (Fort Payne Limestone capped with St. Louis and Warsaw formations).

152.6 Exit 152: Bucksnort. Located along the right of the interstate just a few feet before the exit ramp are exposures of black, nodular chert within a gray limestone of the Mississippian age St. Louis Formation (Fig. 78).

151.3 On your right, notice the St. Louis Formation exposed in the cutslope: again you can see the black nodules of chert in the gray limestone (Fig. 79). The chert, composed of quartz, is very resistant to weathering and can often be found as fragments in the soil.

149.3 Cross Duck River. The Duck River, which flows north of I-40, is joined by the Buffalo River before flowing into the Tennessee River just east of New Johnsonville.

148.5 Exit 148: Turney Center, S.R. 229 interchange. (The Turney Center, part of the state prison system, is located at Only, Tennessee.)

148.1 Leave Hickman County; enter Humphreys County.

144.6 Here you cross the crest of a drainage divide, leaving the drainage area of the Duck River and entering that of the Buffalo River. The hills are underlain by very cherty

Fig. 78. Large black nodules of chert (silica) characterize the Mississippian St. Louis Formation near the Bucksnort exit on I-40, Log Mile 152.6.

Mississippian age strata (Fort Payne Formation), but no bedrock is visible at this point.

143.1 Exit 143: S.R. 13 interchange, Linden, Waverly. (Side Trip 7 to Wells Creek structure and Dunbar Cave State Natural Area begins at this exit.)

142.0 Here you encounter the wide, flat floodplain of the Buffalo River, one of the few "wild" (undammed) rivers left in Tennessee. Canoeists enjoy its scenic course.

141.4 As you cross the Buffalo River, you leave its drainage area and enter that of the Tennessee River. Here the hills are underlain by deeply weathered Fort Payne Formation, which produces a reddish-orange, cherty, clay soil. The hills are capped with another Mississippian age limestone, the Warsaw Formation.

136.9 Exit 137: Cuba Landing.

136.7 Here I-40 crosses the Tennessee River Wildlife Refuge, an area of wetlands (adjacent to the Tennessee River in this vicinity) that provides habitat for birds and other wildlife.

Fig. 79. The horizontal beds of Mississippian-age limestone exposed along I-40 near Log Mile 151.3 are broken by vertical joints (cracks) that enable surface runoff to infiltrate and weather the strata.

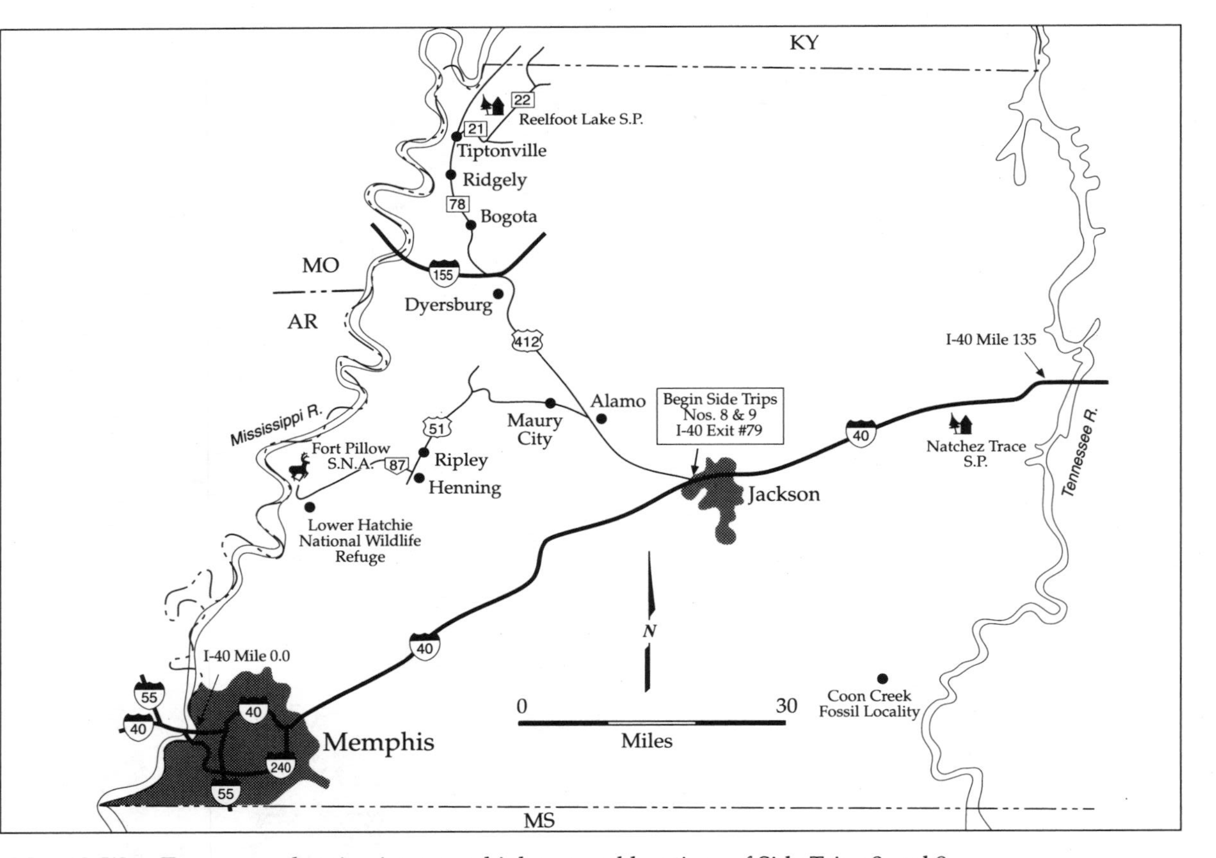

Map 10. West Tennessee, showing interstate highway and locations of Side Trips 8 and 9.

West Tennessee

Mile 135.0 to Mile 0.0 (Mississippi River: Tennessee–Arkansas state line)

Physiographic Provinces

Western Valley of Tennessee River: Miles 135.0–133.0
Gulf Coastal Plain: Miles 133.0–1.1
Mississippi River Alluvial Floodplain: Miles 1.1–0.0

135.0 As you cross the Tennessee River, it is running north (to your right as you travel west on I-40) at this point (Map 10). The river also serves as the line between Benton County, which you are entering, and Humphreys County, which you are leaving.

133.0 Exit 133: S.R. 191, Birdsong Road. Here you leave the western valley of the Tennessee River, which separates the Western Highland Rim from the Gulf Coastal Plain province (Map 11; Fig. 80). (The Devonian age Birdsong Shale member of the Ross Formation, though not visible at this particular location, is exposed in various other spots in the area.)

131.2 Rest area off westbound lane (with restrooms).

130.1 Rest area off eastbound lane (with restrooms).

128.2 At this point you gradually leave an area underlain by the Mississippian age Fort Payne chert and begin traversing a region that is underlain by Cretaceous age sediments. The Cretaceous strata, dipping gently to the west, reflect the deposition of sediments more than 60 million years ago in a shallow sea called the Mississippi Embayment.

127.4 The hill you are crossing is capped with Cretaceous age sand deposits (Coffee Formation). Notice the reddish-orange to brown sandy soil along the left of the interstate.

126.1 Exit 126: S.R. 69, U.S. 641 interchange, Camden, Parsons. At this point the landscape is underlain primarily by the Cretaceous Coffee Sand. Some exposures of weathered Mississippian, Devonian, and Silurian rocks (mostly shale and limestone) can be found south of the interstate, along U.S. 641, toward Parsons.

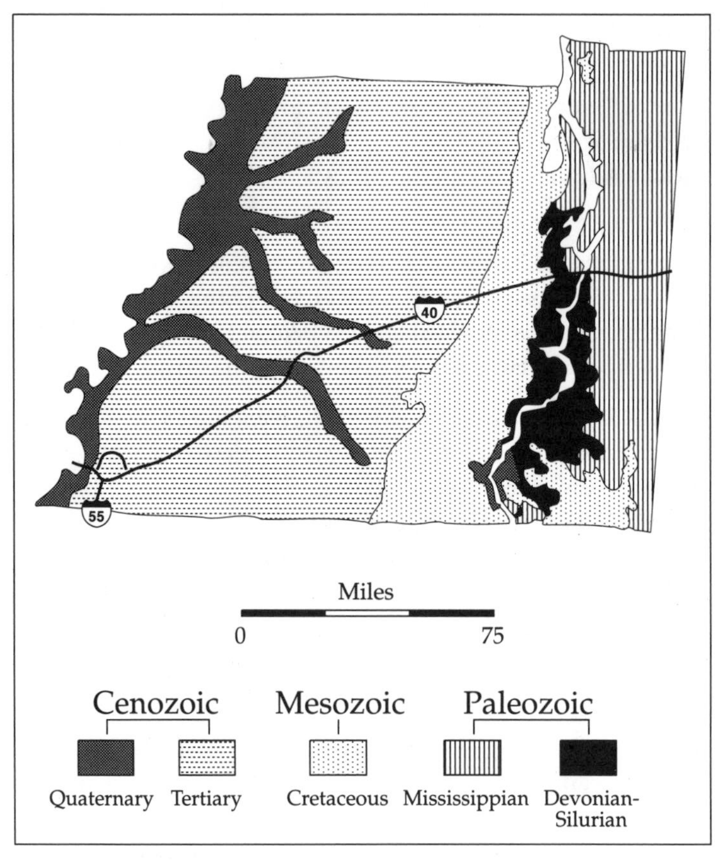

Map 11. Generalized geologic map of West Tennessee. After Geologic Map of Tennessee, Tennessee Division of Geology, 1966.

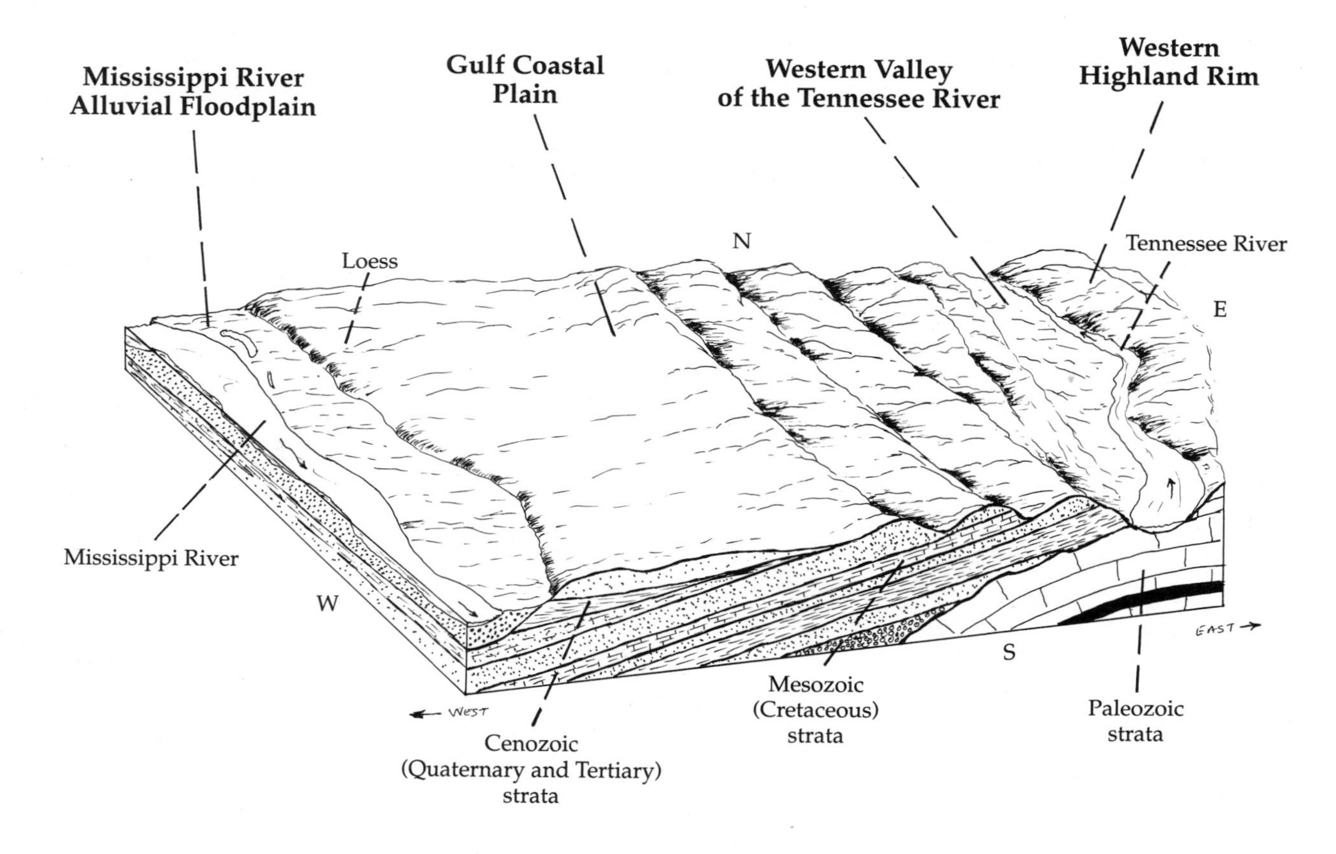

Fig. 80. This schematic block diagram illustrates the geologic structural conditions in West Tennessee.

126.0 Leave Benton County; enter Decatur County.

125.0 Here you cross a small, isolated exposure of the Fort Payne Formation, which you can see along both sides of the interstate. This exposure marks the approximate location of the contact between Cretaceous age sediments (Coffee Sand) with Mississippian age limestone (Fort Payne Formation). The Coffee Formation, consisting of light reddish-brown sand, can be seen along the gravel county road adjacent and running parallel to the right side of I-40.

122.0 Here you begin to encounter the Coon Creek Formation, a greenish-gray, sandy, micaceous marl that caps the highest hills for the next 2.5 miles. As you continue westward, the Coon Creek Formation, dipping to the west, gradually floors the valleys. As you approach Mile 116.5, the highest hills are capped with McNairy Sand, a reddish-orange, sandy soil; the lower slopes of the hills as well as the valley floors are underlain by the Coon Creek Formation.

120.3 Leave Decatur County; enter Carroll County.

119.6 Leave Carroll County; enter Henderson County.

116.6 Exit 116: Natchez Trace State Park, S.R. 114. (You may be interested in visiting this park, which is named for the famous American Indian and pioneer trail running from Nashville, Tennessee, to Natchez, Mississippi. Covering about 43,000 acres, it is the largest of Tennessee's state parks, with opportunities for such activities as camping, fishing, and hiking, and facilities including a small museum and nature center as well as a resort inn and restaurant. The land here, now covered with forests, was badly eroded when the U.S. Department of Agriculture acquired it in the mid-1930s for a "Land Use Area" to demonstrate the benefits of proper conservation practices in rehabilitating heavily eroded areas.)

The hill at the location of the highway interchange and other hills westward for several miles are underlain by McNairy Sand. At this point exposures of the Coon Creek Formation end as you go west, and an area underlain primarily by the McNairy Formation begins. Although exposures of McNairy Sand are not very common along I-40, you can occasionally see them in fields and some eroded banks and cutslopes. In this area almost all land surfaces that are not forested are seeded in grass to prevent erosion of the sandy soil.

116.5 Leave Henderson County; again enter Carroll County.

115.5 Leave Carroll County; again enter Henderson County.

112.8 To the left of the interstate is an eroded bank where the reddish-orange soil of the McNairy Formation is visible.

110.8 The Big Sandy River, which you cross here, parallels the eastern edge of the Tertiary age outcrop area in Tennessee. At this point, the landscape becomes flatter. As you go west, you will notice a sharp decline in pronounced ridges and an increase in flat or rolling topography.

108.2 Exit 108: S.R. 22 intersection. (S.R. 22 leads south to Shiloh National Military Park, the site of the second-greatest battle of the Civil War.)

You can see that both before and after you cross the Big Sandy River the hills are pronounced. These hills are capped with Cretaceous age sand (McNairy Formation).

107.0 Here you leave the widespread area of the McNairy Formation and enter an area underlain by the Tertiary age Midway Group of sand and clay.

104.6 The hills you see to the right of the interstate are capped with the Tertiary age Clayton Formation, an argillaceous sand, but the strata of this formation are not visible from the road. Cretaceous age McNairy Sand is in the valleys. The westernmost exposure of McNairy Sand can be seen along the gravel county road south (to the left) of I-40.

103.6 This is the drainage divide between the Mississippi River to the west and the Tennessee River to the east.

103.0 Parking area off eastbound lane (no restrooms).

102.3 Parking area off westbound lane (no restrooms).

100.9 Exit 101: S.R. 104 interchange.

98.6 Cross Middle Fork of Forked Deer River. Here you begin to climb out of the floodplain of the Forked Deer River, leaving the rolling hills and pastures underlain by the Midway Group.

97.6 This is the approximate point of contact between the brown sand and white clay of the Tertiary age Wilcox Formation and the brown to brownish-gray clay of the Midway Group. The sediments of the Midway Group are thought to be slightly older than those of the Wilcox and Claiborne formations. Unfortunately, there are no notable exposures for you to see along the interstate.

95.0 Leave Henderson County; enter Madison County. Notice the reddish-orange sand in cuts along adjacent county roads and ditches; this is the Tertiary age Wilcox Formation. This hilly section is the drainage divide between the South Fork and the Middle Fork of the Forked Deer River. I-40 follows this drainage divide for the next 15 miles.

93.4 Exit 93: S.R. 152.

86.9 Exit 87: S.R. 1 (U.S. 70) interchange.

82.2 Here you leave the surface exposures of the Tertiary age Clayton and Wilcox formations and begin to encounter the Quaternary age loess deposits. Loess, which is a slightly clayey and sandy silt, originated thousands of years ago during the Pleistocene epoch as outwash sediments from the meltwater of retreating glaciers located to the northwest of Tennessee in what is now the upper Midwest (Iowa, North and South Dakota, Minnesota). Windstorms blowing from the northwest brought the airborne silt to what is now known as West Tennessee, Western Kentucky, and Western Mississippi. The loess material accumulated in thickest amounts along the present route of the Mississippi River (see Side Trip 9 to Fort Pillow State Historic Area) and then eastward for approximately 50 to 60 miles.

Because of the angular shape of the silt particles, the loess soil is more stable when excavated on vertical or near vertical slopes; on flatter slopes it becomes less stable and more erodible. Accordingly, in highway construction where loess is present excavation for roadbeds is usually done on the steep slopes rather than the flatter ones. Material from the flatter slopes is instead used for fill.

At this point you are at the approximate eastern edge of the loess deposits. From here to the Mississippi River at Memphis (about 70 miles), the loess deposits cover the land surface and the Tertiary age sediments beneath. Agriculture, the main industry in West Tennessee, occurs primarily on the loess sediments, although alluvial floodplain sediments, developed where westward-flowing streams cut across the loess deposits, are also used where seasonal flooding does not hamper crop production.

82.0 Exit 82: intersection with U.S. 45. You can occasionally see loess deposits at the surface in this vicinity along I-40.

80.5 Exit 80: Keith Short Bypass (Jackson); S.R. 186 interchange. (Jackson, Tennessee, was the home of the railroad engineer John Luther Jones, better known as Casey Jones. A railroad museum dedicated to him is located there.)

79.2 Exit 79: interchange with S.R. 20 (Hollywood Drive) and U.S. 412. (Side Trip 8 to Reelfoot Lake State Park and Side Trip 9 to Fort Pillow State Historic Area begin here.)

78.9 Leave Jackson City Limits and Urban Boundary.

78.1 As you cross the South Fork of Forked Deer River, notice the wide, flat floodplain. These wetland areas provide habitat for waterfowl and other wildlife.

73.5 Rest area off westbound lane (with restrooms).

72.6 Rest area off eastbound lane (with restrooms).

67.0 Leave Madison County; enter Haywood County.

65.7 Exit 65: S.R. 1 (U.S. 70) interchange.

60.2 Exit 60: S.R. 19 (Old Mercer Road) interchange.

56.3 Exit 56: S.R. 76 interchange.

55.9 As you look west, your view is of the wide, flat bottomland of the Hatchie River (Fig. 81).

Fig. 81. Near Log Mile 55.9, I-40 begins a descent into the flood plain of the Hatchie River (notice the drop in elevation of the path of I-40).

55.3 Now you enter the Hatchie River floodplain, and the interstate descends to the floodplain area.

55.2 Cross Hatchie River. This river flows north out of northern Mississippi into West Tennessee. (An area near the confluence of the Hatchie with the Mississippi River has been designated as the Lower Hatchie National Wildlife Refuge.)

52.3 Here you leave the Hatchie River floodplain. West Tennessee is rich in agricultural products, including cotton (Fig. 82).

47.8 Exit 47: S.R. 179 (Dancyville Road) interchange.

43.1 Leave Haywood County; enter Fayette County.

41.9 Exit 42: S.R. 222 and Stanton Road interchange.

35.2 Exit 35: S.R. 59 interchange.

29.6 Cross Loosahatchie River. The alluvial material deposited in this river and others in the area consists of silt that is mainly redeposited loess washed (eroded) from cultivated fields.

Fig. 82. Numerous crops are produced in West Tennessee, including wheat, soybeans, corn, and cotton. Pictured is a cotton plant in bloom.

27.2 Leave Fayette County; enter Shelby County.

25.3 Exit 25: S.R. 205 and Airline Road; Arlington.

20.5 Exit 20: Canada Road.

18.2 Enter Memphis Urban Boundary.

18.1 Exit 18: S.R. 15, U.S. 64 interchange.

16.9 Exit 16: Germantown Road interchange; S.R. 177.

14.0 Exit 14: Whitten Road interchange.

12.2 Exit 12: Sycamore View Road interchange.

11.1 Wolf River.

10.8 Interchange of I-40 and I-240. *Note:* If you are going to visit the Pink Palace Museum in Memphis, continue straight ahead; do not exit to I-240. The I-40 roadway here changes to Sam Cooper Boulevard. Continue on Sam Cooper Boulevard and take the Highland Street exit. Turn left onto Highland Street and follow it to Central Avenue (four blocks). Turn right on Central; go about one mile, and the museum is on the right. The Pink Palace Museum houses an excellent collection of Cretaceous age fossils from the Coon Creek Formation in McNairy County, Tennessee. The name of the museum comes from the color of the marble used for the building's exterior facing.

Please note that mileage numbers change as you enter I-240 North from I-40 West. On I-40 you are at Mile 10.8 and join I-240 at Mile 11.5 on I-240. Interstate 240 is a loop highway around Memphis.

Take the exit to I-240 North, which you follow 11.5 miles along the north side of Memphis, rejoining I-40 at Mile 2.8 on I-40 (Mile 0.0 on I-240) on the west side of the city. During the 1960s and 1970s concerned citizens waged a successful campaign to block the construction of I-40 through Overton Park, a beautiful wooded area in the heart of Memphis. As a result, I-40 now ends at Mile 10.8 and detours around Memphis via I-240. The shortest route back to I-40 heading west is to follow I-240 around to the north. The extra distance by taking I-240 adds 3.5 miles to the total log mile distance of I-40 across Tennessee (455.33 miles); the total milepost figure, measured as if I-40 were constructed straight through Overton Park to the Mississippi River, is 451.8 miles.

Memphis, known as the Bluff City, was built on what is referred to as the Fourth Chickasaw Bluff, overlooking the Mississippi River (Parks and Lounsbury 1975: 36). This bluff consists of loess deposits (up to 65 feet thick), which are underlain by other Quaternary gravel deposits and the Tertiary clay, sand, and lignite of the Jackson Formation and Claiborne Group.

Memphis is one of the largest cities in the United States to depend solely on groundwater for its water supplies (Parks and Lounsbury 1975: 42). Most of the groundwater is contained in two sand aquifers (water-bearing strata of permeable material) above 2,000 feet in depth. Called the Memphis Sand and the Fort Pillow Sand, these two aquifers are Tertiary in age.

2.8 Interchange with I-40. Exit I-240 onto I-40 West (Mile on I-40 is 2.8).

1.4 Exit 1: S.R. 1.

1.1 Now you leave the Gulf Coastal Plain province and enter the Mississippi River Alluvial Floodplain province. "Old Man River," the Mississippi, is very close to the bluffs here.

0.6 Cross old channel of Wolf River. Riverside Drive exit. (Take this exit for a short excursion while you are in the Memphis area: a visit to Mud Island Park. This historical park and museum on an island in the Mississippi River has an outdoor scale model of the Mississippi River showing the border between Tennessee and Arkansas with all the river's meanders, islands, and oxbow lakes. Mud Island Park also has an intriguing museum about the history of Memphis and the influence of the Mississippi River on the city. After taking the Riverside Drive exit, follow the signs to the park.)

Fig. 83. In Memphis, I-40 crosses the Mississippi River, which forms most of the western boundary of Tennessee. Located near the I-40 bridge over the Mississippi is Mud Island.

0.5 Cross Mud Island (Fig. 83).

0.0 Mississippi River: Tennessee-Arkansas state line. The Mississippi River floodplain extends well into Arkansas but is very limited on the Tennessee side. The loess bluffs at Memphis abruptly end the floodplain.

Side Trips

1. Cumberland Gap National Historic Park

Destination: Pinnacle Overlook, Cumberland Gap National Historic Park (Kentucky, Virginia, Tennessee)
Route: I-81, U.S. 25E (includes S.R. 32, S.R. 1, and S.R. 33), and Park Service road; begins at I-40, Exit 421 (Map 12)
Cities and towns: Morristown, Bean Station, Tazewell, Harrogate, Cumberland Gap
Trip length: 64.8 miles one-way (to Pinnacle Overlook); to return to I-40, retrace the route
Nature of the roads: Four-lane, divided interstate; four-lane sections of primary roads; hilly and curvy sections of two-lane, paved highways; Park Service road is two-lane, paved road, very hilly, steep, and curvy
Beginning elevation: 1,290 feet
Ending elevation: 2,440 feet
Special features: Traverse of Valley and Ridge province, Clinch Mountain, Paleozoic rock strata, engineering geology, tilted sedimentary strata
Hiking trail (optional): Pinnacle Overlook
U.S.G.S. 7 1/2 minute quadrangle maps: Middlesboro 153 NW and 153 SW
For reference: Valley and Ridge province, Paleozoic era, caves, fossils

Directions. From I-40 take Exit 421 onto I-81 North; go north on I-81 to Exit 8 (approximately 8 miles) and exit onto U.S. 25E north. Follow U.S.

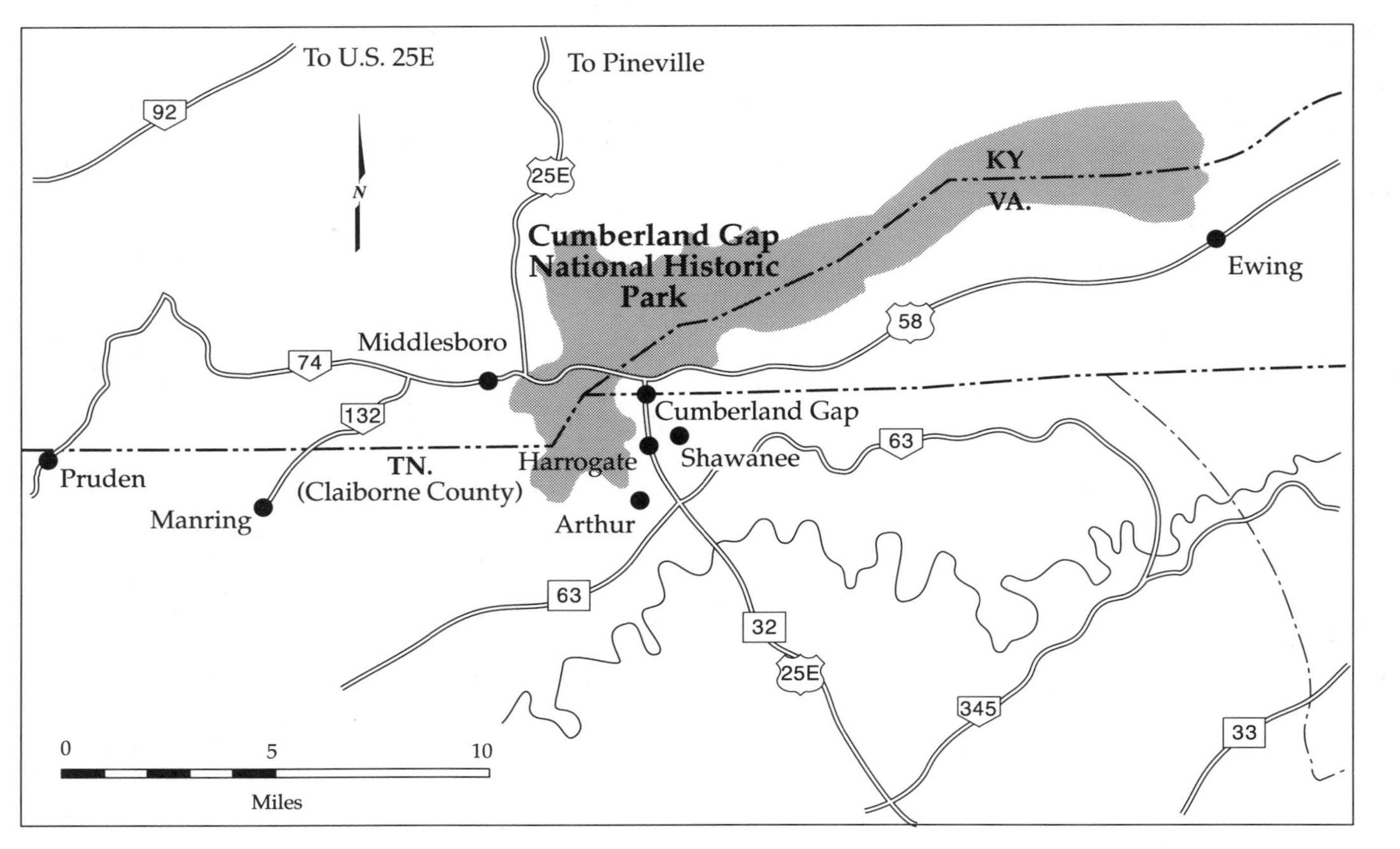

Map 12. Location map of Cumberland Gap National Historical Park.

25E across Hamblen, Grainger, and Claiborne counties (approximately 50.2 miles). Continue on U.S. 25E north across the tip of Virginia (0.8 mile) and across Cumberland Gap into Kentucky (1.6 miles) to the exit for the Cumberland Gap National Historic Park Visitors Center. Follow the exit ramp approximately 0.3 mile to the Visitors Center. To travel to the Pinnacle Overlook, turn right from the Visitors Center parking area onto the paved park road and follow the signs up the mountain for 3.5 miles to the Pinnacle Overlook parking area.

Note: Although the Pinnacle Overlook road is paved, it is very steep and contains many sharp, switchback turns. Accordingly, this road is not recommended for large vehicles (camper trailers, etc.) or during inclement weather.

Route details. Along this route you will see the rolling landscape of the Valley and Ridge province, climb atop Clinch Mountain, cross the drainage areas of three rivers (Holston, Clinch, and Powell), and traverse up the slopes of Pine Mountain to Pinnacle Overlook. This route takes you into three states: Tennessee, Virginia, and Kentucky.

As you begin this trip, you will be crossing a relatively rolling section of landscape underlain by tilted shale and limestone strata in Jefferson and Hamblen counties (Log Mile 0.0 to Log Mile 17.6). After passing by Walters State Community College and the city of Morristown, you will cross the first of three rivers, the Holston River at the Hamblen-Grainger county line (Log Mile 17.6) on U.S. 25E; S.R. 32 in Hamblen County. About 9 miles into Grainger County (near Log Mile 26.5), you will begin to cross very hilly and mountainous sections of the Valley and Ridge province. Clinch Mountain, which punctuates this section of the route, takes you up to 2,000 feet in elevation and provides scenic views to both the north and the south (Fig. 84).

The construction of U.S. 25E across Clinch Mountain was a major undertaking in terms of engineering geology. The approximately six-mile traverse up and down the mountain cost about $17 million. Two areas of geological problems were encountered, both involving types of translational landslides: block-glide landslides on the south slopes of the mountain and colluvial landslides along the north scarp side (Aycock 1981).

Block-glide (planar-failure) landslides are areas of movement of earth and bedrock along a flat surface (plane), which is usually inclined 15° to 30° from the horizontal. Most often the movement of block-glide landslides occurs along a bedding plane where clay may have developed

Fig. 84. From the top of Clinch Mountain you are able to see the parallel Valley and Ridge topography. To the east the large body of water is Cherokee Lake.

between two layers of rock (Fig. 85). Because excavations by highway construction activity along the south side of Clinch Mountain removed the support of the base of the tilted layers of rock, large areas of bedrock slid.

Along the north side of Clinch Mountain are deposits of unstable broken rock debris and soil (colluvium) that accumulate along the surface of the lower slopes of the mountain. When these colluvial deposits are disturbed (either by excavation or by placement of soil or rock on top of the colluvium), landsliding usually takes place. Such areas of movement—characteristically in a straight, downslope direction, with the disturbed material moving as a whole mass rather than in sections—are typical of translational landslides.

Rock buttresses were designed to correct the landslide problem along the south slopes of the mountain. These retaining structures, constructed out of blasted rock, appear as large masses of white and light brown blocks of rock stacked in a somewhat orderly fashion along the roadway cut slopes. An unusual adaptation of the rock buttress concept, known as a shot-in-place rock buttress, was first used on this stretch of highway: in this method the sliding rock mass is drilled and blasted into a disorganized mass of rock debris (Fig. 86; Aycock 1981:

Fig. 85. An ancient landslide scar on Clinch Mountain near Log Mile 30.5. You can easily see the dipping beds of strata of the Clinch Sandstone.

19; Royster 1979). This process breaks up the planes along which sliding is taking place and provides a mass that will restrain the unstable material upslope. The resulting slope appears as an unevenly excavated cutslope with some rock and soil debris exposed (Fig. 87). Such slopes are usually well vegetated. These rock buttresses and slide areas are located along the right of the highway from Log Mile 28.0 to Log Mile 30.5 (Fig. 88).

Along the north side of the mountain, a technique called underbenching was used to provide stability for the highway embankments on which you now ride. Underbenching was devised by highway engineering geologists as a method for dealing with problems encountered in constructing highway embankments (fills) on steeply sloping, colluvium-covered terrain. Flat, horizontal benches, which are excavated in the sloping land surface down to bedrock, provide a solid foundation for the construction work. As the highway roadbed is constructed, the underbenches are covered up with soil and rock. Although you cannot see the underbench areas now, they are located beneath the highway embankment from Log Mile 31.4 to near Log Mile 33.0.

This stretch of highway across Clinch Mountain and north to the Clinch River is known by geologists as the Thorn Hill geologic section

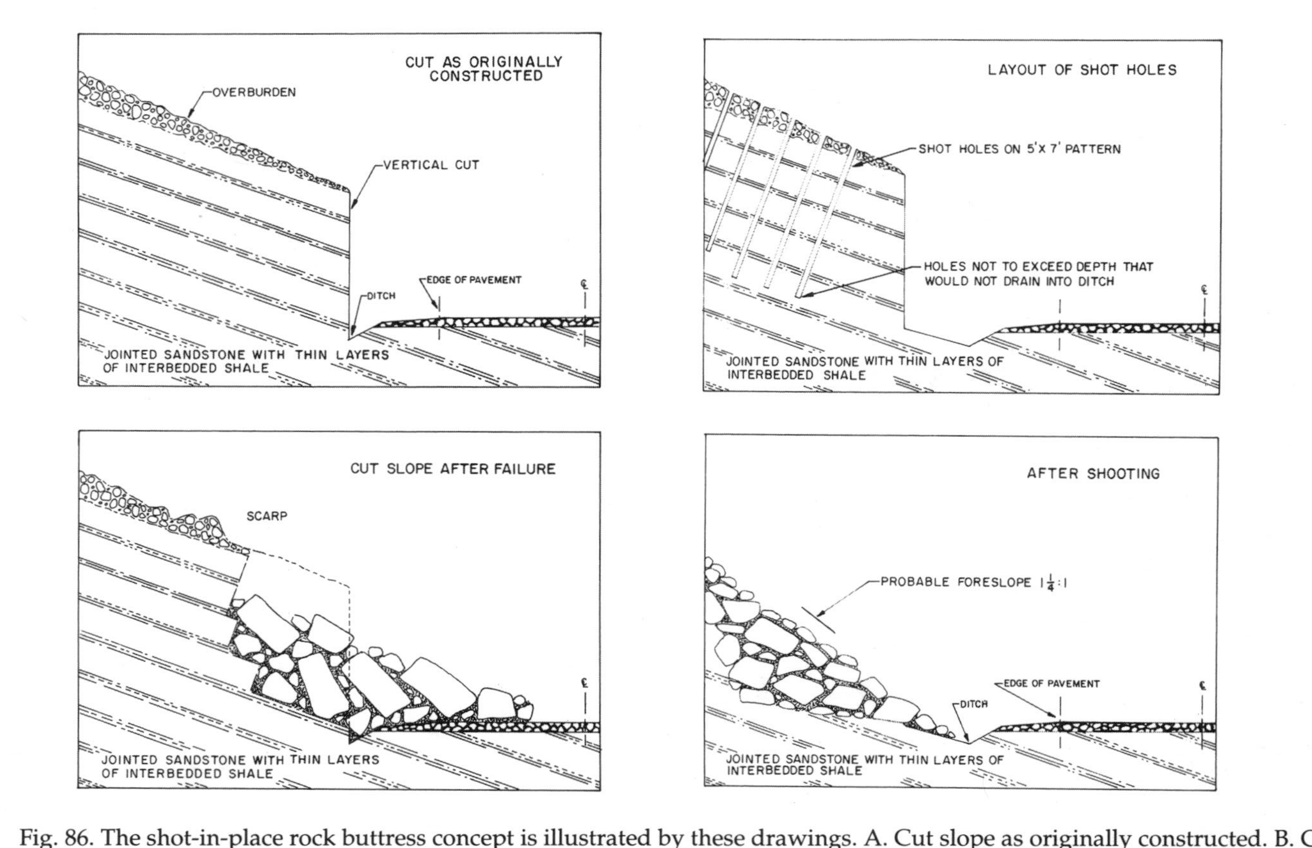

Fig. 86. The shot-in-place rock buttress concept is illustrated by these drawings. A. Cut slope as originally constructed. B. Cut slope after landslide. C. Redesigned cut slope with location of drill blast holes for shot-in-place buttress. D. Rock slope after blasting the rock slope and reshaping the surface forming the buttress.

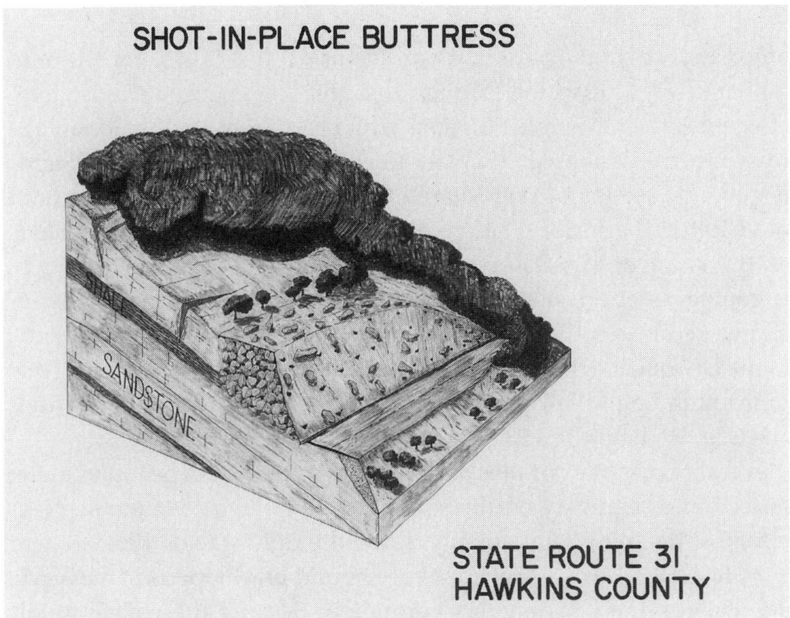

Fig. 87. These schematic block diagrams illustrate a typical block-glide landslide (top) and shot-in-place rock buttress (bottom) used on S.R. 31 on Clinch Mountain in Hawkins County.

Fig. 88. This rock buttress is one of many on the south slope of Clinch Mountain. This buttress was used to restrain a block-glide landslide.

(named for the small community on the north slopes of Clinch Mountain). Highway excavation through this area has exposed a most complete geologic stratigraphic section, with rocks from the Cambrian age through the Mississippian (K. R. Walker et al. 1985). Your route begins this geologic section at Log Mile 27.0, where the Saltville thrust fault (not visible at the surface) has displaced the Cambrian age Rome Shale onto the younger Mississippian sandstone of the Grainger Formation. The geologic section continues along the highway, from the Mississippian age sediments (Log Mile 27.0–27.5), down through geologic time into the Devonian age Chattanooga Shale (the black rock from Log Mile 27.5 through Log Mile 28.3), and on into the Silurian age Clinch Sandstone (Fig. 89; Log Miles 28.3–31.3).

Several excellent examples of invertebrate fossils can be found along this section of highway on the south slope of Clinch Mountain. Near the base of the mountain (notably, along the right side of the road near Log Mile 28.0), there are specimens of lingulid brachiopods in the black, platy shale of the Chattanooga Formation. Because these shell fossils are very small (about the size of a small fingernail), you will need a hand lens to find them. Specimens of the fossils *Arthrophycus* and *Scolithus* are easy to find in the light brown to grayish sandstone strata

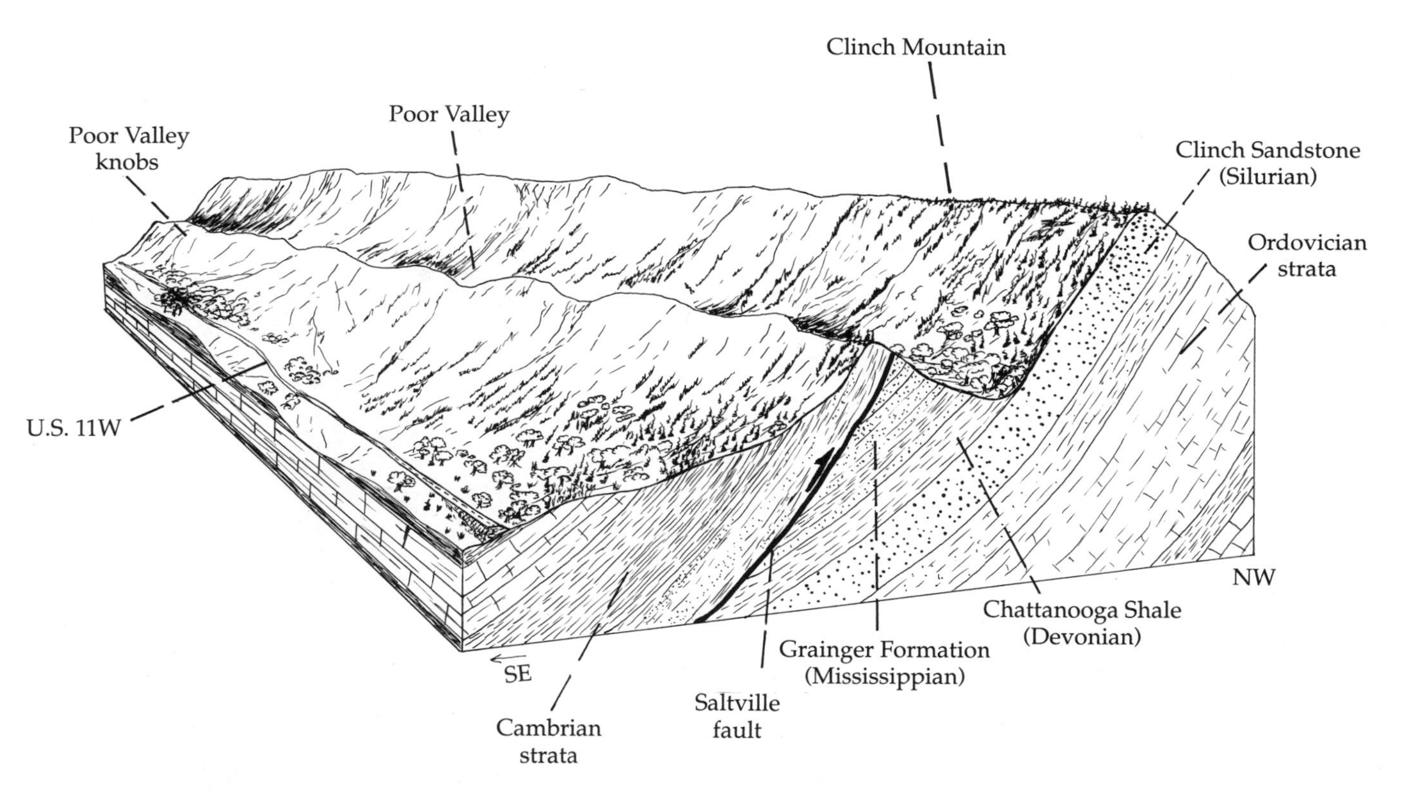

Fig. 89. Illustrated in this schematic block diagram is the geologic structure found along the Clinch Mountain section of Side Trip 1 (Log Mile 30.5).

of the Clinch Formation. *Arthrophycus,* the preserved horizontal borings of a marine worm (Fig. 90), can be seen on the flat surfaces of the many blocks of sandstone lying along the right side of the road (Log Miles 28.3–31.3) as you continue up the south slope of Clinch Mountain; look for the characteristic segmented, arching, worm-like patterns. The vertical, pencil-shaped lines also present in many of these pieces of the Clinch Sandstone are fossils known as *Scolithus,* believed to be a vertical burrow of a marine organism.

There, at the top of the mountain, the geologic section continues into the Ordovician age, where brown and maroon shale and gray and pink limestone are found down the north side of Clinch Mountain and into the adjacent Clinch Valley (Log Miles 31.3–34.0). Near Log Mile 34.0 the Ordovician limestone gradually yields to Cambrian age gray limestone and greenish-gray to brown shale. At Log Mile 35.4 to the right of the highway is a limestone quarry where dark gray to black limestone of the Cambrian age Conasauga Group is quarried and polished for building stone.

The Cambrian age strata end abruptly at Log Mile 35.8, where the Copper Creek fault (not clearly visible) has thrust the Cambrian age Rome Shale (maroon, green, gray, and brown) onto Ordovician age

Fig. 90. The Silurian-age Clinch Sandstone contains numerous specimens of the fossil *Arthrophycus,* which is believed to be the preserved horizontal borings of a marine worm.

Fig. 91. Located on the right side of the road near Log Mile 36.0 is an excellent example of a thrust fault. The fault plane is the dark diagonal band that dips from the upper left to the lower right. The rocks to the right of the fault have been thrust over the rocks to the left of the fault.

limestone and shale. The Copper Creek fault marks the lower boundary of the complete Thorn Hill geologic section, from Mississippian age strata at Log Mile 27.0 to lower Cambrian age strata at Log Mile 35.8. Just past the Copper Creek fault, a small but very distinct thrust fault is exposed in the rock slope along the right of the road at Log Mile 36.0 (Fig. 91). This smaller fault is thought to be associated with the Copper Creek thrust fault.

At the Grainger–Claiborne county line (Log Mile 37.7), the route takes you over the Clinch River, a beautiful wild and scenic river that flows into Norris Lake, the first of many lakes created by TVA dam projects in the Tennessee Valley. Just north of Tazewell the route passes over another river, the Powell (Log Mile 53.4).

Finally, you come to Harrogate, Tennessee, home of Lincoln Memorial University. Located on the LMU campus is the Abraham Lincoln Museum, one of the nation's most complete collections of materials dealing with Lincoln and the Civil War (over 25,000 items). After passing through Harrogate (and bypassing the tiny town of Cumberland Gap), you will travel for a short distance into Virginia, where the existing U.S. 25E passes in front of the entrance to Cudjo's Caverns. Then you go into Kentucky, across Pine Mountain and Cumberland Gap.

The National Park Service has plans to purchase the caverns from the private owner and incorporate the caverns into the park area, providing an interpretive service for visitors to the cave. The cave acquisition is in response to a future change in the location of U.S. 25E when the twin tunnels now under construction carry traffic through the mountain from Tennessee into Kentucky. The old highway route across the mountain will be permanently closed when the new tunnels are finally open.

Along this route you have traveled in two physiographic provinces and three states, crossed three rivers and two mountains, and seen numerous rock formations. You have now reached a place where you can see three states from one vantage point: Pinnacle Overlook.

General Geology of Cumberland Gap

The Cumberland Gap area is underlain by Paleozoic sedimentary strata consisting of limestone, shale, siltstone, sandstone, and coal beds of both marine and coastal swamp environments (Fig. 92; Vandover 1989: 3–6). After the deposition and accumulation of these sediments, tremendous forces (resulting from the colliding North American and African continents some 250 million years ago) acted upon these strata, deforming the once horizontal layers into structural folds. These stresses were so great that older rocks from the southeast were thrust for miles over younger rock strata to the northwest. One of these thrust sheets, known as the Pine Mountain overthrust, includes this section of Pine Mountain stretching from Tennessee into Kentucky.

Subsequent weathering and erosion of the strata have resulted in the landscape you see now. The parallel ridges and valleys in East Tennessee and Virginia are underlain by folded and faulted strata. The high mountain areas of eastern Kentucky and adjacent portions of East Tennessee are underlain by the Pine Mountain overthrust, which consists of slightly tilted strata of Pennsylvanian age.

Cumberland Gap itself is only a wind gap along the crest of Cumberland Mountain. It is believed that former stream activity in the gap area cut a notch or saddle into the very resistant but highly fractured strata. More easily eroded areas to the northwest (in and around Middlesboro) "robbed" the stream activity, diverting the flow into what is now known as the Cumberland River. Continued erosion developed the water divide on the crest of the ridge, and the former course of the stream became the gap. Water that now falls on the southeast side of the Cumberland

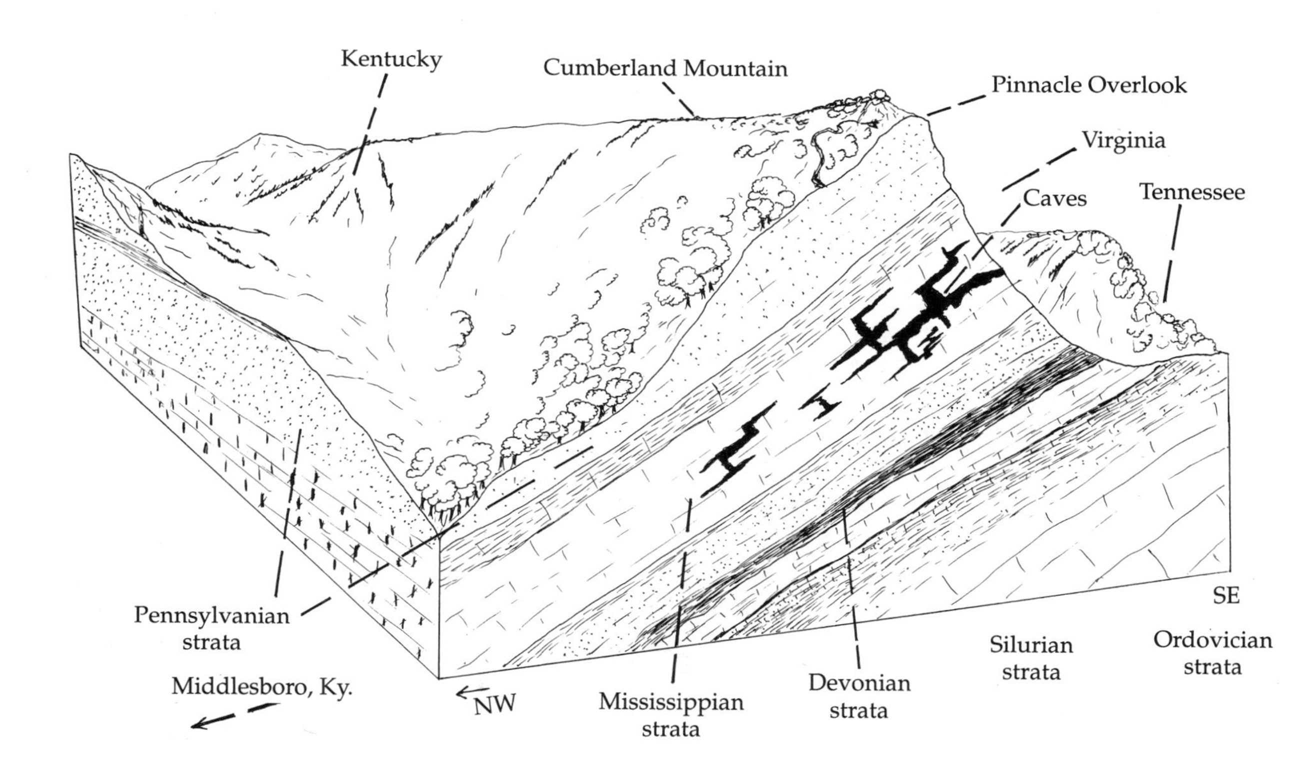

Fig. 92. This geologic block diagram illustrates the subsurface conditions at Cumberland Gap National Historical Park. Note the solution cavities in the Mississippian-age limestone. Cudjo Cave is located in these rocks.

Mountain ridge crest flows into the Tennessee River, while water falling on the northwest side of the crest flows into the Cumberland.

Cumberland Gap Tunnel Project

In an effort to restore the gap to its 1790–1800 appearance and to remedy a dangerous and heavily traveled section of U.S. Highway 25E, the Federal Highway Administration is constructing a highway tunnel through Cumberland Mountain. Two parallel tunnels, each 4,100 feet long (one for northbound traffic and one for southbound), are being built through the mountain and will connect to new highway approach sections in Tennessee and Kentucky.

A pilot bore, ten feet in diameter, was completed through the tunnel section in 1986 to aid contractors in their bidding assessments on the contracts for the main tunnel bore. In May 1991 the contracts were initiated with the excavation of the tunnel bores. Completion of the tunnel project, which will involve excavating many thousands of tons of rock, is estimated for 1995.

The new tunnels will cut through dipping sedimentary rock strata that range in age from Devonian to Pennsylvanian. The actual bore cuts perpendicular to the strike (or trend) of the dipping beds of strata and the resulting cross-section slices through the rock, showing its dipping structure. When the tunnel is completed, however, it will be lined with concrete and the bedrock will not be visible from within the tunnel. The lining is for prevention of falling rocks in the tunnel as well as for aesthetic reasons.

The tunnel construction has encountered Pennsylvanian age sandstone, shale, and coal seams along the Kentucky sections. Near the midpoint of the tunnel, a cavern was encountered; developed in Mississippian limestone, this cavern has produced large amounts of clear spring water. Local speleologists, attempting to map the cavern, have surveyed over a mile of passageway.

Near the Tennessee portal, or entrance to the tunnel, the presence of shale and thin-bedded, jointed Devonian and Silurian age limestone made it necessary to provide special steel and concrete supports for the portal. (Stabilizing broken and weathered rock formations near the ground surface is a common problem around most tunnel entrances.) Although the tunnel itself is not visible, you can see the new approach roads leading to the tunnel entrance on the Tennessee side of the mountain from the Pinnacle Overlook.

Pinnacle Overlook Trail (optional)

Destination: Pinnacle Overlook
Location: Trail begins at Pinnacle Overlook parking area at end of paved Park Service road
Start: Pinnacle Overlook parking area, elevation 2,440 feet
End: Pinnacle Overlook, elevation ± 2,420 feet
Nature of the trail: Paved foot trail, some steps
Distance: 400 yards, round-trip
Special features: Spectacular view into three states—Virginia, Tennessee, and Kentucky; views of the Valley and Ridge and Cumberland Plateau provinces; exposures of Pennsylvanian sandstone

Historical significance. Native Americans were the first humans to use the gap, establishing it as a major route into the hunting grounds of Kentucky. It became an important feature of the Warrior's Path, a trail connecting the Ohio River with the Potomac. White men probably first entered the gap in 1750—a result of survey work by Thomas Walker. After wars with the French and Indians, peace again flourished in the area. In 1775, Daniel Boone and thirty men marked the wilderness trail from Cumberland Gap into Kentucky that was later used by thousands to enter the new territory, the American West. By the 1820s and 1830s, however, when the West became easily reached by steamboats, the Pennsylvania main line, or the Ohio and Chesapeake Canals, the usefulness of the gap as a transportation route diminished somewhat (Vaughn 1927: 1–11).

During the Civil War the gap was considered strategically crucial for holding the surrounding territory. At first the Confederates held the gap, but Union troops captured it on June 17, 1862. Three months later the Confederates retook it when the Union army evacuated the area. Cumberland Gap was finally captured and held by Union troops on September 9, 1863 (Vaughn 1927: 1–11). Today the gap is still a major transportation route, with a railroad (via tunnel) and federal highway traversing it.

It is important to note that early settlements were established in response to travel routes and the economic opportunities afforded by natural resources. Geology, of course, is crucial to both. The first settlers in the Cumberland Gap area found low-grade iron ore in the Silurian age rocks (Rockwood Formation) along the base of the mountain below the gap on the Tennessee side. Ample supplies of coal from the adjacent Pennsylvanian age coal deposits in Kentucky and limestone from Tennessee provided the necessary materials for smelting iron ore.

In Cumberland Gap the presence of the Newlee Iron Furnace attests to the mineral resources of the area. Built about 1815, the furnace smelted the ore into iron, which could then be used to make many useful items, such as axes, plow points, and guns. At the time when the furnace was operating, smoke and fumes from the smelting process probably filled the air around Cumberland Gap. The furnace, long inactive, is now part of the Cumberland Gap National Historic Park, adjacent to the town of Cumberland Gap, Tennessee, and is open daily for touring.

Another nearby site of historic interest is Hensley Settlement, a restored mountain community with 12 homesteads which lasted from its beginning in 1904 to the early 1950s. The settlement prospered, attaining a peak population of about 100 in the decade of 1925 to 1935. Abandoned in the early 1950s, Hensley Settlement deteriorated until 1965, when the Park Service began a restoration program (Beatty 1978: 20–23). It can be reached by the Ridge Trail, mentioned in "Additional Hiking," below.

The trail is short and the reward handsome. The paved trail begins in Kentucky at the upper end of the parking area, next to a sheltered area housing a sculpture. After a short flight of steps, the trail quickly curves around a high hill, entering the state of Virginia, and descends a ramp-like pathway to Pinnacle Overlook, which juts outward.

Trail description. On the way to the overlook, the trail passes an outcrop of cross-bedded Pennsylvanian age sandstone that you can easily see to the left of the trail, just before you reach the overlook. Look for a grayish to light brown rounded rock with thin layers that curve at different angles.

The wide-angle view from the pinnacle is spectacular (Fig. 93). Starting on the right, you see the state of Kentucky, with the city of Middlesboro in clear view. As your gaze moves to the left, you will see the deeply forested slopes of Pine Mountain (part of the heavily eroded Cumberland Plateau province). Straight ahead of you the landscape changes to the Valley and Ridge topography of Tennessee, with broad, green, pastured valleys and forested ridges. The small town of Cumberland Gap can be seen, nestled in the valley below, as well as the twin highway bridges connecting the new tunnel to the adjacent ridge.

The view to the left reveals a large bluff composed of Pennsylvanian age sandstone. This is Virginia! The sandstone is medium to light gray and contains the same cross-bed structures you saw in the sandstone outcrop along the trail on your way up to the pinnacle. The grains of sand in this rock are composed of the mineral quartz. Because quartz is very resistant to weathering forces, the sandstone forms the top layers of the mountain.

Beneath the pinnacle is a thick sequence of Mississippian age limestone containing the passageways of Cudjo's Caverns. Groundwater has dissolved some of the limestone bedrock, forming numerous interconnecting passageways that correspond to cracks (joints) in the limestone. Although the passageways in Cudjo's Caverns are in the limestone beneath the pinnacle, they are inaccessible here. To get to the entrance of the cave, you must return to U.S. 25E (Log Mile 58.6).

As you return to the parking area, a short loop trail of 400 yards, which will take you back to the trailhead, begins on the right. This trail (steep, with numerous steps) will lead you to the earthwork ruins of Fort Lyon (Civil War) at the crest of the hill.

Fig. 93. From the Pinnacle Overlook at Cumberland Gap National Historical Park you can see the mountain through which the new U.S. 25E roadway tunnel passes. The two white horizontal lines to the left of the mountain are the new roadway bridges leading to the tunnel on the Tennessee side.

Additional hiking. If you would like a longer hike, there are several trails that can be recommended. Beginning at the Pinnacle Overlook parking area is Ridge Trail (16 miles), which follows the crest of the mountain past Hensley Settlement to White Rocks. This trail, going along the boundary between Kentucky and Virginia, offers many scenic views into both states. Hikers on this trail need to carry their own water, as this is a dry trail. The Ridge Trail makes connections with seven other trails, including routes to Skylight Cave, Sugar Run Trail, Gibson Gap Trail, Hensley Settlement, Shillalah Creek Trail, Chadwell Gap Trail, and Sand Cave.

The Skylight Cave Trail begins at Lowes Hollow picnic area and proceeds 0.9 mile (one-way) to the cave area. The Tri-State Trail, which leads to the point where Tennessee, Kentucky, and Virginia meet, is also 0.9 mile (one-way).

For additional information on trail locations and descriptions of the scenic Cumberland Gap National Historic Park area, you may wish to consult a useful trail guide to the park by Steven M. Beatty, *Why Not Walk?* (Eastern National Park Association, 1978), or Robert S. Brandt, *Tennessee Hiking Guide,* rev. ed. (Knoxville: Univ. of Tennessee Press, 1988).

Additional information (food, lodging, schedules of events):

Superintendent, Cumberland Gap National Historic Park
P.O. Box 1848
Middlesboro, KY 40965
(606) 248-2817

Abraham Lincoln Museum
Harrogate, TN 37752-0901
(615) 869-3611

Claiborne County Chamber of Commerce
P.O. Box 332
Tazewell, TN 37879
(615) 626-4149

Grainger County Chamber of Commerce
P.O. Box 101
Rutledge, TN 37861

Middlesboro Tourism Commission
City Hall Building
P.O. Box 756
Middlesboro, KY 40965
(606) 248-5670

Morristown Area Chamber of Commerce
P.O. Box 9, 825 West First North Street
Morristown, TN 37815
(615) 586-6382

Lee-Norton-Wise-Scott
Planning District Commission
P.O. Box 366
Duffield, VA 24244
(703) 431-2206

2. Cades Cove, Great Smoky Mountains National Park

Destination: Cades Cove, Great Smoky Mountains National Park
Route: U.S. 321 (S.R. 95 and S.R. 73) and Park Service road; begins at I-40, Exit 364 (Maps 13 and 14)
Cities and towns: Lenoir City, Maryville, Walland, and Townsend
Trip length: 54.6 miles one-way to Cades Cove; Cades Cove loop road is an additional 11 miles
Nature of the roads: Four-lane, divided highways; some two-lane, paved highways with curvy and hilly sections; Park Service road, two lanes and paved, is very curvy and sometimes closed in winter because of snow and ice; Cades Cove loop is one-lane, paved road, very narrow and curvy with hilly sections, and very slow driving
Beginning elevation: 837 feet
Ending elevation: 1,964+ feet
Special features: Cades Cove and Tuckaleechee Cove geologic windows, tilted Paleozoic and Precambrian age sedimentary rocks, scenic views of the Great Smoky Mountains (some peaks over 6,000 feet in elevation)
Hiking trail (optional): Gregory Bald
U.S.G.S. 7 1/2 minute quadrangle maps: Cades Cove 148 SE
For reference: Paleozoic, Precambrian, Valley and Ridge province, Blue Ridge province, coves, caves, colluvium, faults, folds, cleavage, bedding, Cambrian, Ordovician

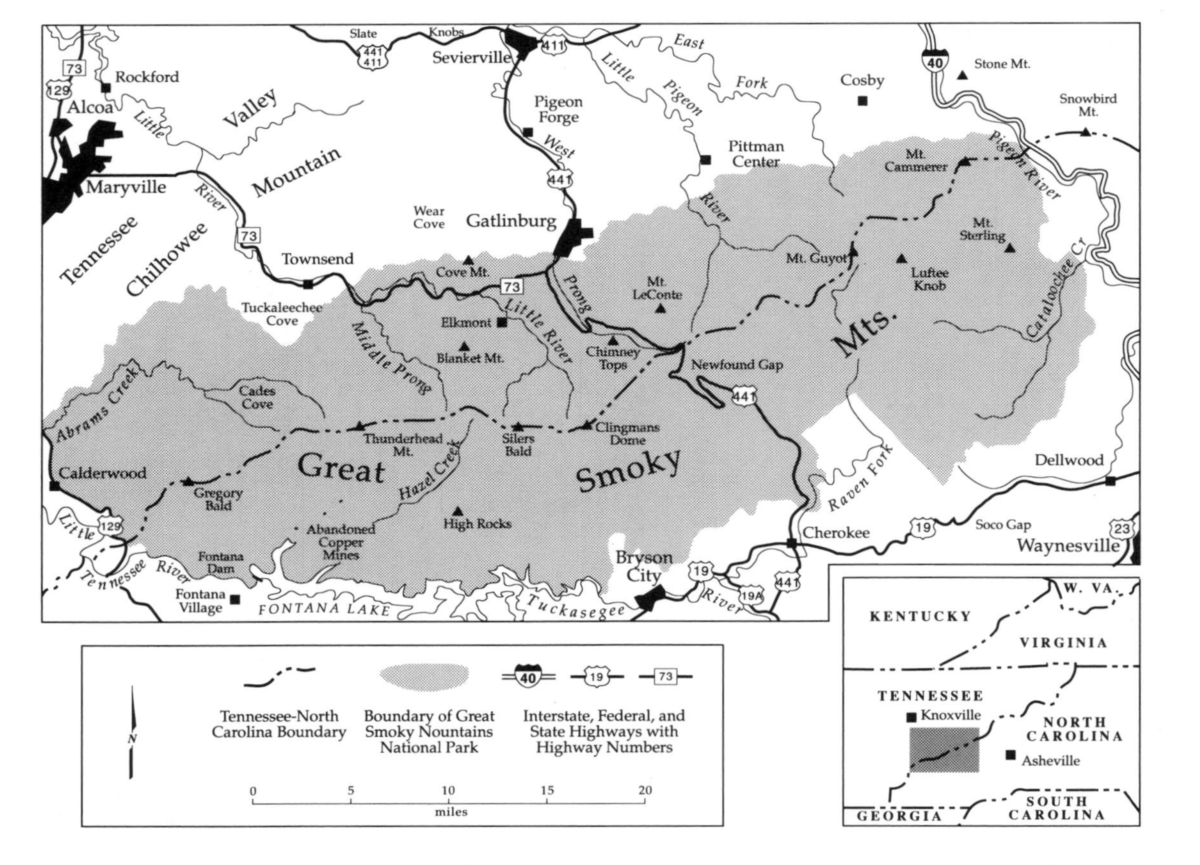

Map 13. A generalized location map of Great Smoky Mountains National Park. Base map after P. B. King 1949, on back of U.S.G.S. Map, "The Great Smoky Mountains National Park and Vicinity" (revised 1972).

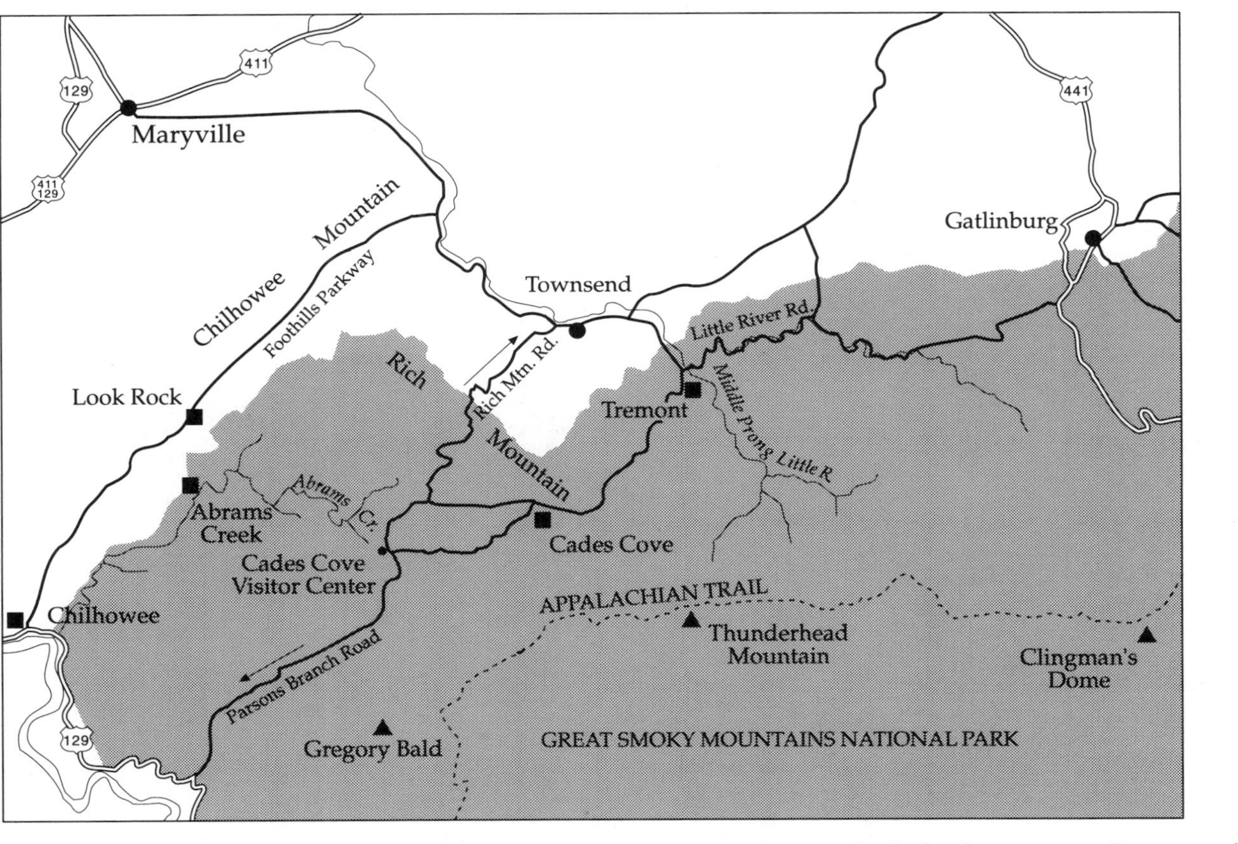

Map 14. Northwest section of Great Smoky Mountains National Park, showing the Cades Cove vicinity. Base map after U.S. Dept. of Interior, National Park Service map, "Great Smoky Mountains National Park, North Carolina/Tennessee."

Directions. From I-40 take Exit 364 onto U.S. 321. Turn left (south) onto U.S. 321 and follow it for 44.5 miles, passing through Lenoir City, Maryville, Walland, and Townsend. In Townsend, U.S. 321 turns left to Pigeon Forge. Do not take this left turn. Instead, take S.R. 73, which continues straight ahead. About 1.5 miles past the U.S. 321 turnoff is the entrance to the Great Smoky Mountains National Park. Enter the park and continue for 0.9 mile. At the fork in the road, you veer to the right. Now you are on Laurel Fork Road. Follow this park road for 7.7 miles to Cades Cove and the beginning of the one-way loop road.

Route details. Along this route you will have views of the rolling Valley and Ridge topography, mountainous Blue Ridge landscapes, and scenic mountain coves. Your destination, Cades Cove, is one of the most beautiful and geologically fascinating areas in the Great Smoky Mountains National Park.

The route begins along a section of rolling landscape in the Valley and Ridge province that is underlain by Cambrian and Ordovician age limestone and dolostone strata. Weathering has reduced surface exposures to spot occurrences; most of the area is covered by residual clay soil. Roadway excavations in this clay soil are designed on 2:1 ratios (about 27°) because the internal stability of such soil is not sufficient to permit steeper slopes.

At Log Mile 3.8 the route intersects U.S. Highway 70 (Eaton Crossroads), and at Log Mile 4.6 the route passes over Interstate 75. At Log Mile 7.3 the side trip route intersects U.S. Highway 11 in Lenoir City. After passing through the commercial districts of Lenoir City and crossing U.S. 11, the route enters the TVA's Fort Loudon Dam reservation at Log Mile 7.8 (Fig. 94). The highway passes directly over the top of Fort Loudon Dam at Log Mile 8.3, where you can see the waters of Fort Loudon Lake upstream and the Tennessee River downstream. Fort Loudon Dam, constructed between 1940 and 1943, is the uppermost dam on the Tennessee River containing navigable locks.

A short distance downstream (one-half to one mile), the Little Tennessee River, after being dammed by the Tellico Dam, joins the Tennessee River. The construction of the Tellico Dam (1967 to 1979) was halted for several years by the discovery of the snail darter, a small, minnow-sized fish that was placed on the U.S. Endangered Species list.

For the next 3 to 4 miles the route traverses the drainage divide between the Tennessee River and the Little Tennessee. At Log Mile 12.8,

Fig. 94. The route of Side Trip 2 passes directly over the TVA's Fort Loudon Dam at Log Mile 7.8.

S.R. 95 turns south (to the right). Stay on U.S. 321. At Log Mile 15.0 you leave Loudon County and enter Blount County. Tilted Cambrian and Ordovician strata are found throughout this section of Blount County. You can see some notable rock outcrops of the Lenoir Formation (Ordovician limestone) on the right side of the road from Log Mile 17.4 through Log Mile 17.6. At Log Mile 19.0 the outcrops are Holston Formation (also Ordovician limestone): this rock exposure is found along the side of a small hill along the right side of the roadway, off the right-of-way, in a small quarry pit that was developed to obtain rock for the roadway construction.

Near Log Mile 22.7 the route crosses a high, broad ridge underlain by the Knox Group of strata. On clear days you can see the mountains of the Blue Ridge province in the distance; Chilhowee Mountain is the most prominent peak.

From Log Mile 24.0 through Log Mile 30.0, you will pass through the commercial district of Maryville, Tennessee. Points of interest include the Blount County Courthouse (c. 1907) at Log Mile 27.2, Maryville College (established in 1819) at Log Mile 27.5, and the historic Thompson-Brown House (c. 1800) at Log Mile 27.8.

Located adjacent to Maryville in the town of Alcoa is a large plant of the Aluminum Company of America (Alcoa). Even though the route does not pass directly by the plant, Alcoa's presence is worth mentioning because it is related to the TVA dams you are seeing. Bauxite, the ore of aluminum, is shipped by rail to this area, where it is smelted and refined into aluminum. Alcoa chose to build its plant in this area because of the abundant source of cheap hydroelectrical power.

The route continues across tilted Cambrian and Ordovician strata (mostly weathered to residual clay soil). At Log Mile 33.0 there is a dramatic view of Chilhowee Mountain, which marks the boundary between the Valley and Ridge and the Blue Ridge provinces.

Rock outcrops along the roadway cut section begin to show increased deformation (tightly folded, tilted zones of extreme crushing and faulting) near Log Mile 36.0 (Fig. 95). This deformation is the result of faulting activity from the Great Smoky thrust fault, a major fault that moved older, highly deformed, and altered Precambrian rocks over younger Cambrian and Ordovician strata. Some researchers estimate that these rocks were moved by the action of the fault from as far away as possibly 200 miles to the southeast. The movement of the rock strata by thrust faulting is the result of compressional forces produced by the gradual collision of the North American continent with the African continent some 250 million years ago. It was this convergence of forces that produced the Appalachian mountain-building episode (Alleghanian orogeny).

At Log Mile 36.2 you will see the beginning of an unusual-looking concrete retaining wall along the right side of the road. This is what is known as a tie-back wall: hundreds of steel anchors, 60 to 80 feet long, were drilled into the mountain in order to hold the concrete slabs in position to stabilize the side of the mountain. The checkerboard appearance of the wall is a result of the construction method (Fig. 96).

A trace of the Great Smoky thrust fault is exposed at Log Mile 36.25 near the beginning of the sandstone exposure beneath the tie-back wall. The fault is exposed where the brown to brownish-gray sandstone (Cambrian, Hesse Formation) comes in contact with the dark gray shale material (Ordovician, Sevier Formation). The highly contorted, folded, and deformed gray shale (Sevier Formation) just west of the retaining wall has been documented to contain numerous fossil graptolite species and possibly fish scales (Hatcher and Lemiszki 1991).

Traveling southeast toward the Great Smoky Mountains, you pass through Miller Cove, where rocks of the Cambrian age Chilhowee Group (Nebo Sandstone, Murray Shale, and Hesse Sandstone) and

Fig. 95. Near Log Mile 36.0 you can see highly deformed rock strata—a result of movement along the Great Smoky Thrust Fault.

Fig. 96. The Tennessee Department of Transportation constructed this tie-back retaining wall instead of making a large cut into the side of Chilhowee Mountain (Log Mile 36.2).

Shady Formation (a dolostone) are exposed. You can see the dolostone strata, medium to light gray in color and medium bedded, to the right of the highway at Log Mile 37.6, just past the interchange with the Foothills Parkway (Log Mile 37.3).

At Log Mile 39.2 the roadway cut has exposed tightly folded slate and siltstone of the Precambrian Wilhite Formation. The rock is varied in color, from brownish-gray to rusty brown, and the folds are chevron shaped, 30 to 40 feet in height and width. To observe these beautiful structures, you must park just southeast of the cut, where a gravel road enters the highway, and walk about 100 feet back to the exposure. *Caution:* Be mindful of the highway traffic.

The route continues through rocks of the Precambrian Walden Creek Group (mostly Wilhite Formation), which are folded and contain cleavage structures. Notable rock exposures are found at Log Miles 40.2 and 40.6. Near Log Mile 41.4 the highway enters Tuckaleechee Cove (Townsend), crossing the Great Smoky thrust fault, where erosion has cut down through the older, overlying Precambrian rocks, exposing the more soluble limestone and shale of Paleozoic age.

Due to the soluble nature of the limestone, numerous caves are found in Tuckaleechee Cove. The commercialized Tuckaleechee Caverns are the most widely known of the caves in the Townsend area (the turnoff to Tuckaleechee Caverns is at Log Mile 42.9).

Beginning at Log Mile 43.0 and continuing for about 3 miles is a paved bicycle trail on the right side of the highway. At Log Mile 44.5, U.S. 321 turns left to Pigeon Forge; you should continue straight on S.R. 73 for 1.5 miles to the entrance to the Great Smoky Mountains National Park.

At Log Mile 45.0 on the way toward the park entrance, you will pass by Cedar Bluff, a huge, gray bluff easily visible to the left of the highway across the Little River. If you carefully observe the thick-bedded strata of Cedar Bluff (limestone and dolostone of the Paleozoic Knox Group), you will find several faults exposed in the bluff wall. As you look at the bluff, you can see the dipping layers of rock that abruptly change their orientation, forming a linear surface that separates the rock layers. This exposure of rock with its associated structure was fundamental for giving geologists an understanding of the movement of the Great Smoky fault. The fault is best observed during the winter months, when there is less vegetation covering the face of the bluff.

At Log Mile 45.8, approximately 1.3 miles past the U.S. 321 turnoff, is an exposure of rock on the right side of the road that shows the structure of the Great Smoky fault. Late Precambrian phyllite (Metcalf Phyllite of the Snowbird Group) is thrust over Paleozoic limestone and shale with a slice of the limestone present above the other Paleozoic rocks. This is a horse structure—a large block of displaced wall rock (limestone in this case) that is caught along a fault surface.

The route continues on into the Great Smoky Mountains National Park, and you soon approach the Y intersection about 0.9 mile from the entrance. Turn right onto Laurel Creek Road and drive to Cades Cove. Along most of this section of road, you will notice numerous exposures of a shaly rock, medium to light gray: this is the Precambrian age Metcalf Phyllite. The thin, platy layering you see in these strata is not bedding but foliation and cleavage resulting from the structural deformation induced by faulting and folding. Along the first mile of roadway, you will also see exposures of the Precambrian Cades Sandstone, large beds of brownish-gray sandstone.

Approximately 6.7 miles beyond the "Y" the route crosses Crib Gap (Log Mile 53.6), where you begin descending into Cades Cove. Cades

Cove is another geologic "window," a structure where you can look down through older strata to see the younger strata. In this case, erosion has penetrated down through older Precambrian rocks, which are thrust over younger Paleozoic limestone and shale. The erosion has penetrated the fault plane (the Great Smoky fault) and exposed the underlying Paleozoic strata (limestone and shale). As a result, the valley is surrounded by ridges and mountains composed of older, faulted strata, while the valley floor is underlain by younger limestone and shale (Fig. 97).

At Log Mile 54.4 (approximately 0.8 mile from Crib Gap), a road turns left from the route and leads to the Cades Cove ranger station, picnic areas, camping areas, and a small camp store. Continuing straight ahead for about 0.25 mile on the route, you will reach the beginning of the 11-mile, one-way Cades Cove Loop Road at Log Mile 54.6. This scenic drive is a must for experiencing the flavor of the cove.

In Cades Cove you get a glimpse into the history of a unique mountain community (Fig. 98). Originally the cove was inhabited by the Cherokee. Then, in the fall of 1818, the John Oliver family moved from Carter County in the northeast corner of East Tennessee and settled in the cove. Other settlers followed, clearing parts of the cove for farming and the construction of houses, barns, churches, and a gristmill (Dunn 1988: 35).

Rural life continued in the cove until the late 1920s, when land began to be bought up for the new Great Smoky Mountains National Park (to be established officially in 1934). Many of the inhabitants of the cove—including John W. Oliver, a great-grandson of the original settler, who held a 350-acre tract—were unwilling to sell, and the Government resorted to condemnation proceedings (Campbell 1960: 98–99; Dunn 1988: 241).

By the mid-1930s most of the families had left the cove, and the park was faced with a dilemma. Once the farmers had gone, the forest began to reclaim the area, encroaching on the scenic views of the surrounding mountains. Although the usual park policy was to let the land go back to nature, in this case the solution was to make Cades Cove a historical area within the park and thus maintain the pastures as part of the setting (Campbell 1960: 147).

In this scenic drive you will have the opportunity to see not only log houses, split-rail fences, and open pastures, but also some of the wildlife of the area, such as white-tailed deer, wild turkeys, skunks, groundhogs, and an occasional black bear. The historic Cable Mill area (with a visitors center and restrooms) is about halfway around the loop road.

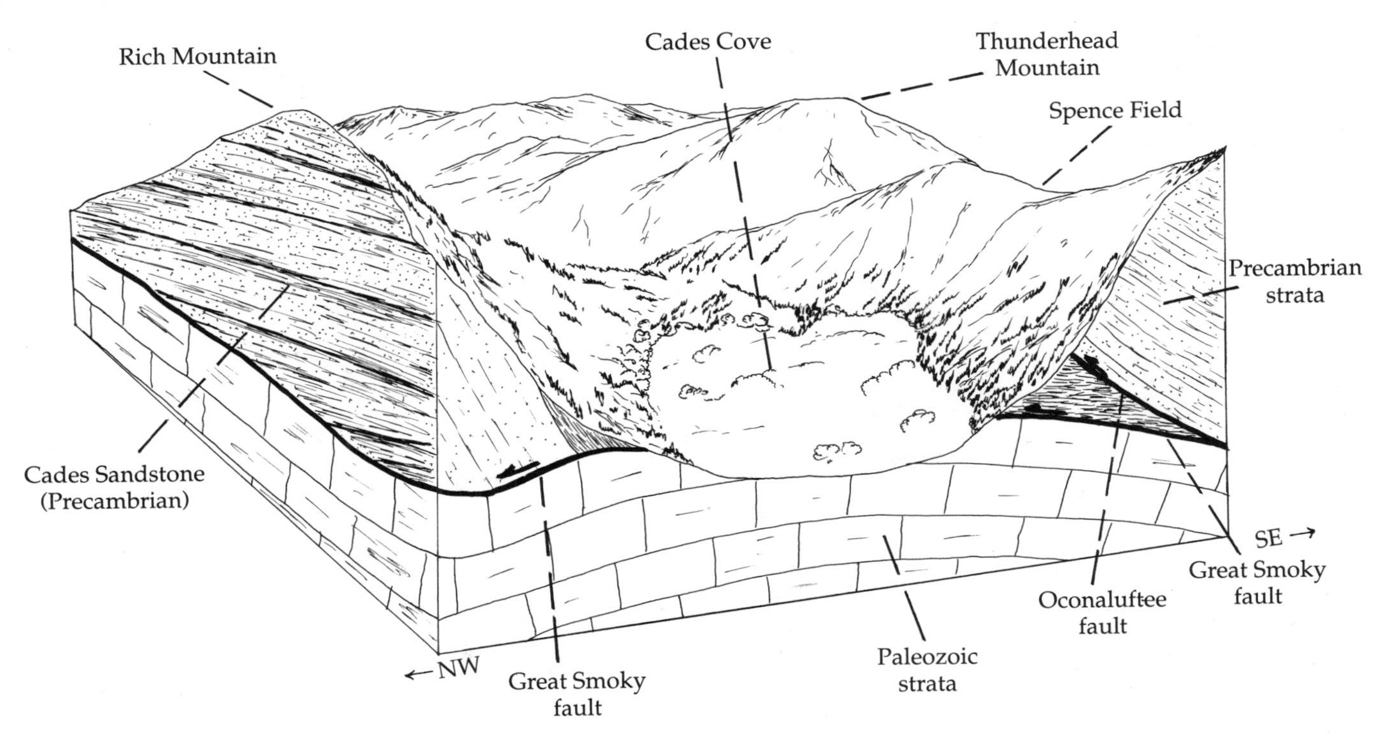

Fig. 97. Cades Cove was formed when erosion penetrated the Great Smoky Thrust Fault (which lies almost horizontal) to expose Paleozoic limestone and shale.

Fig. 98. Cades Cove, in the heart of the Smoky Mountains, was home to early settlers of the area.

The valley surface of Cades Cove is mostly covered by colluvial deposits representing several periods of extensive erosion and deposition of debris from the mountain slopes. Most of the colluvium, which consists of boulders and transported soil debris, is found along the ravine drainages and the lower "toe" areas (that is, the bases) of the surrounding slopes. Research in this area indicates that the colluvium may have originated as debris flows and/or mudflows, reaching thicknesses of as much as 300 feet, with the most recent colluvial activity occurring 2,800–4,000 years before the present (Clark et al. 1989; Davidson, 1983). Following is an abbreviated road log around the paved Cades Cove Loop Road.

Mileage	Description
0.0	Begin 11-mile scenic loop road.
0.4	Irregular, mounded topography may be colluvial mudflow or debris flow.
0.9	Sparks Lane on the left.

1.0 Parking for John Oliver Cabin. The Olivers bought this site and settled here in 1826. The Park Service believes that the cabin, a hand-hewn log structure, was built in the mid-1850s.

1.3 The light gray, rounded masses of rock to the right of the road are outcrops of Paleozoic limestone.

1.9 The Park Service maintenance road on the right leads to a karst area. Among the numerous limestone outcrops is the entrance to Gregorys Cave (Fig. 99). The cave is gated to protect park visitors from injury; for entry into this or any other cave in the park, you must have permission from the Park Superintendent.

2.5 The Methodist church on the right, which dates from 1902, was constructed by one man in 115 days.

2.8 Hyatt Lane (on the left).

3.0 Continue straight ahead. (Rich Mountain Road, on the right, is a one-way gravel road that goes up the south slope of

Fig. 99. Beneath the surface soils of Cades Cove is limestone strata that has undergone extreme weathering to produce cavernous strata. This is a view of the entrance to Gregorys Cave in Cades Cove. Note the limestone strata exposed around the cave entrance.

Cades Cove, over Rich Mountain, and back down into Tuckaleechee Cove.) Missionary Baptist Church is on your left.

3.6 Scenic overlook on the left. From this vantage point you can see the width of the cove and the sloping ridges that have funneled colluvial debris into the cove.

5.0 Cross Abrams Creek. This stream is responsible for the reworking of the colluvium and residual soil of the cove and the continued downward erosion of the valley floor.

5.1 Gravel road to Abrams Falls trailhead on the right. (A detailed description of the hiking trail to Abrams Falls appears in Moore 1988: 139–46.)

5.5 Lane to Cable Mill on the right. You may enjoy leaving the road trip for a short visit to this historically interesting area, where you can see not only the gristmill but also a blacksmith's shop, a cantilevered barn, the Gregg-Cable House (c. 1879), and numerous outbuildings. There are also a visitors center and restrooms. You can easily return to the trip route.

At this point the paved one-way loop road turns left. The gravel road continuing straight ahead is Forge Creek Road. It leads 2.1 miles to the one-way, gravel Parsons Branch Road, which takes you over mountainous terrain to U.S. Highway 129 at the southwest boundary of the park (Fig. 100). This is also the route to the optional hiking trail to Gregory Bald described at the end of this road log.

Continue on the loop road to the left from Cable Mill Lane. This part of the loop traverses mostly colluvium (rounded boulders and rock debris), the result of thousands of years of erosional processes that have produced debris flows into the valley floor.

6.7 Hyatt Lane on the left. On the right is the Dan Lawson Cabin, built in 1856, which is unusual for having a chimney of brick rather than stone. The bricks were made from clay obtained from a nearby creekbank. This clay, in turn, probably originated from the weathering of the limestone in this area to produce a soil rich in the clay mineral kaolinite, which can be fired to make bricks.

Fig. 100. The one-way Parsons Branch Road leads from Cades Cove to U.S. 129 at the west park boundary. To take the hike to Gregorys Bald you can travel this road to Sams Gap to the trailhead.

7.8 Tipton Place. You will observe many colluvial boulders ahead along the road.

8.3 Carter Shields Cabin on the right.

8.7 Sparks Lane on the left. Notice the numerous grayish to brownish, rounded colluvial boulders on both sides of the road.

10.2 End loop road. Turn left and go 0.3 mile to Laurel Creek Road. To return to I-40, turn right onto Laurel Creek Road and retrace the route.

Gregory Bald Hiking Trail (optional)

Destination: Gregory Bald

Location: Trail begins at Sams Gap, 3.3 miles on Parsons Branch Road (Fig. 100). To reach Parsons Branch Road, follow the Cades Cove Loop Road for 5.5 miles to the Cable Mill turnoff; there Forge Creek Road (a gravel road) leads straight ahead for 2.1 miles to where Parsons Branch Road begins on the right. Look for the sign. *Note:* Parsons Branch Road is a narrow, one-way gravel road that leads out of the southwestern end of Cades Cove. Thus, once you commit your vehicle to Parsons Branch Road, you cannot return to Cades Cove by car except by going 8 miles further to U.S. 129 and then back through Walland and Townsend. There are no services along Parsons Branch Road, and it is closed during the winter. When you reach U.S. 129, turn right and follow it for 10 miles to the intersection with the Foothills Parkway (on the right) along Chilhowee Lake. Turn onto the Foothills Parkway and follow it for approximately 17.1 miles to U.S. 321 at Walland. (This is a very scenic drive. For a geologic description of the parkway route, see Moore 1988: 147–55.)

Start: Sams Gap, on Parsons Branch Road; elevation 2,780 feet

End: Gregory Bald, elevation 4,948 feet

Nature of the trail: Unpaved foot trail, moderately difficult; long uphill climb; terrain is mountainous and vegetated

Distance: 8 miles, round-trip

Special features: Spectacular scenic views from Gregory Bald; good example of Blue Ridge topography (but few rock exposures); famous display of wild flame azalea (second half of June);

This hiking trail takes you to a very scenic spot, and, if hiked in late June, provides you with an unforgettable display of wild flame azaleas. The 8-mile round-trip hike leads you to one of the anomalous grassy balds of the high Smoky Mountain crest. When the azaleas are in bloom, the bald radiates an orange and red glow from the thousands of flowers.

Gregory Bald is named after the Gregory family who lived in Cades Cove during the 1800s (Murlless and Stallings 1973: 292). History tells of farmers who used the grassy bald area to graze their cattle and sheep during the summer season (Murlless and Stallings 1973: 292; Dunn 1988: 43–44). The origin of the grassy balds is still debated but is believed to have no relationship to the bedrock (King 1964: 141). The most probable cause of the deforestation of the high ridges is thought to be fire originating from either lightening strikes or early Native Americans (Dunn 1988: 33; Murlless and Stallings 1973: 33).

Several bald areas occur along the crest of the Smokies from Clingmans Dome westward to Gregory Bald. These include Silers Bald, Spence Field, and Russell Field east of Gregory Bald, and Parson Bald west of Gregory Bald.

Trail description. The trail begins across Parsons Branch Road from the parking area at Sams Gap. At first the trail climbs gradually through a pine forest. In the Sams Gap area you cross the east-west fault trace of the Oconaluftee fault, which brings the older, Precambrian Elkmont Sandstone into an overlying contact with the younger, Precambrian Cades Sandstone. The Parsons Branch Road approximates and somewhat parallels the location of the fault.

The trail quickly steepens, climbing out of the pine forest and gradually entering a mature hardwood forest near Panther Gap, at 2.7 miles. From Panther Gap the trail continues steep, climbing Hannah Mountain. The forest above Panther Gap consists mainly of oaks, maples, hickories, and tulip poplars.

The resistant nature of the metamorphosed sandstone of the Elkmont Formation causes the topography to steepen. The numerous sandstone cobbles and rock slabs you notice along the trail in this section are weathered pieces of Elkmont Sandstone.

Near the upper regions of the hardwood forest, you may notice the gray and twisted skeletons of several of the forest ancestors, the American chestnut. Once quite numerous throughout the Smokies, the American chestnut was completely devastated by a parasitic fungus accidentally introduced in New York in 1904. By 1940 almost all of the American chestnut trees in the park were dead (Murlless and Stallings 1973: 33–34). These skeletons still exude an air of majesty as they lie among the ferns, echoing times past, their light gray, weathered wood blending with the shadows and deep green of the underbrush.

At 3.2 miles the trail crosses many water seeps. A short distance later (at 3.5 miles) you reach Sheep Pen Gap, where a backcountry campsite is located. The trail to Gregory Bald turns left (the trail on the right leads 0.8 mile to Parson Bald).

From Sheep Pen Gap to Gregory Bald the trail winds through low trees, mountain laurel, and azaleas. Gradually the trail breaks out of the low forest into the azalea and grass bald. At 4 miles the trail brings you to the grassy crest of Gregory Bald. Here you will see the interplay of flame azaleas with the fine-textured grass.

Near the crest, along the trail, you will notice several low outcrops of steeply dipping dark gray to blackish rock strata. These are metamorphosed rocks consisting of slate and metasandstone. Displacement along the fault moved the rock strata many miles from the southeast to this point.

On clear days views from Gregory Bald include Fontana Lake to the south and the Cheoah Mountains and Snowbird Mountains beyond the lake (all in North Carolina). You can also get excellent views of the western end of the park, which offer opportunities to see the rise of the mountain slopes around Cades Cove to the north, in Tennessee.

When visited in late June, Gregory Bald boasts one of the most beautiful displays of wild flame azaleas in the United States (Fig. 101). The flowers range in color from yellow to orange and pink to red, and many variations in between. The Park Service recently instituted a program whereby park personnel reclaim the balds by cutting back the trees and woody shrubs from the border areas. Scientific evaluations over time will determine the effectiveness of this activity.

For additional information on locations and descriptions of trails in the Great Smoky Mountains National Park, you may wish to consult the useful trail guide to the park by Carson Brewer, *Hiking in the Great Smokies,* 14th ed. (Norris, Tenn.: Carson Brewer); *Scenic Drives and Wildflower Walks in the Great Smokies,* by Betsey B. Creekmore and Betsey Creekmore (Knoxville: Knoxville Garden Club, 1990), provides a good general introduction to the park as well as special emphasis on its plants. For a geologic road log to the park, with optional hiking trails, see Harry Moore, *A Roadside Guide to the Geology of the Great Smoky Mountains National Park* (Knoxville: Univ. of Tennessee Press, 1988).

Additional information (food, lodging, schedules of events):

Blount County Chamber of Commerce
309 South Washington Street
Maryville, TN 37801
(615) 983-2241

Fig. 101. During the last two weeks of June the flowers of wild flame azaleas transform Gregorys Bald into a high mountain flower garden.

Loudon County Chamber of Commerce
P.O. Box K
Loudon, TN 37771
(615) 458-0267

Townsend Visitors Center
7906 East Lamar Alexander Parkway
Townsend, TN 37882
(615) 448-6134

Superintendent, Great Smoky Mountains National Park
Gatlinburg, TN 37738
(615) 436-5615

3. Twin Arches, Big South Fork National River and Recreation Area

Destination: Twin Arches, Big South Fork National River and Recreation Area
Route: U.S. 127, S.R. 154, and Park Service road; begins at I-40, Exit 317 (Maps 15 and 16)
Cities and towns: Crossville, Jamestown
Trip length: 50.6 miles one-way; to return to I-40, retrace the route
Nature of the roads: Two-lane, paved highways, curvy with some straight sections, some hills; Park Service road is narrow gravel road with some hills
Beginning elevation: 1,778 feet
Ending elevation: 1,510 feet
Special features: Natural rock arches, cliffs, and overhangs; plateau landscape; Pennsylvanian age sandstone; flat-lying sedimentary rocks
Hiking trail (optional): Twin Arches
U.S.G.S. 7 1/2 minute quadrangle maps: Sharp Place 335 SE
For reference: Plateau, Pennsylvanian age, Rockcastle and Fentress formations

Directions. From I-40 take Exit 317 onto U.S. 127. Go north on U.S. 127 for 32.5 miles, passing through Jamestown. Turn right onto S.R. 154 (north). At Log Mile 42.7 is Sharps Place (intersection with S.R. 297, which leads to the Big South Fork National River and Recreation Area Visitors Center). Continue north on S.R. 154. At Log Mile 44.7 turn right onto the gravel Park Service road at the sign to Twin Arches. Follow this gravel road for 0.9 mile, then turn left on Divide Road (gravel) and follow it for 3.5 miles. At the fork in the road turn right onto Twin Arches Road, then drive 2 miles to the parking area and trailhead for Twin Arches.

Traveling along this route, you will see the flat to rolling landscape along the top of the Cumberland Plateau; travel near historic spots, such as the town of Rugby and the Alvin C. York Memorial Gristmill; and see the natural beauty of the water-carved canyon of the Big South Fork of the Cumberland River.

Natural Resources of the Cumberland Plateau

The first inhabitants of the Cumberland Plateau region were Paleo-Indians, who appeared there during the last Ice Age. Hunters and gatherers, they used the many rock overhangs, rock shelters, and bluff areas as campsites (Coleman and Smith 1993).

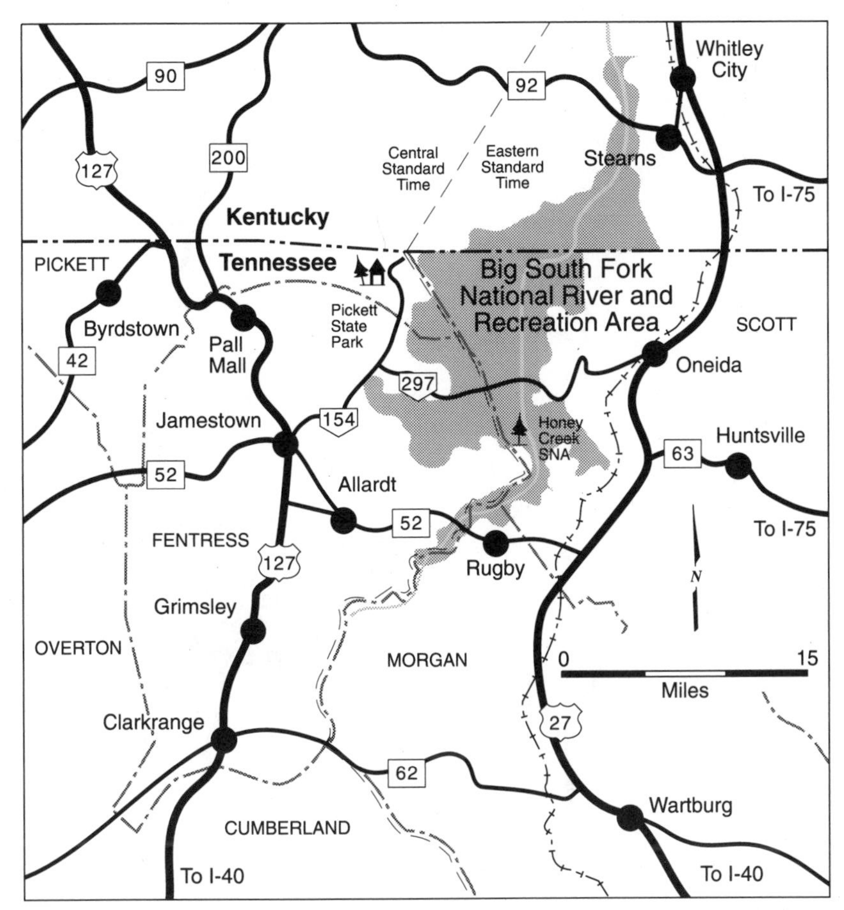

Map 15. Location map of Big South fork National River and Recreation Area. Base map after TDOT Official Highway Map.

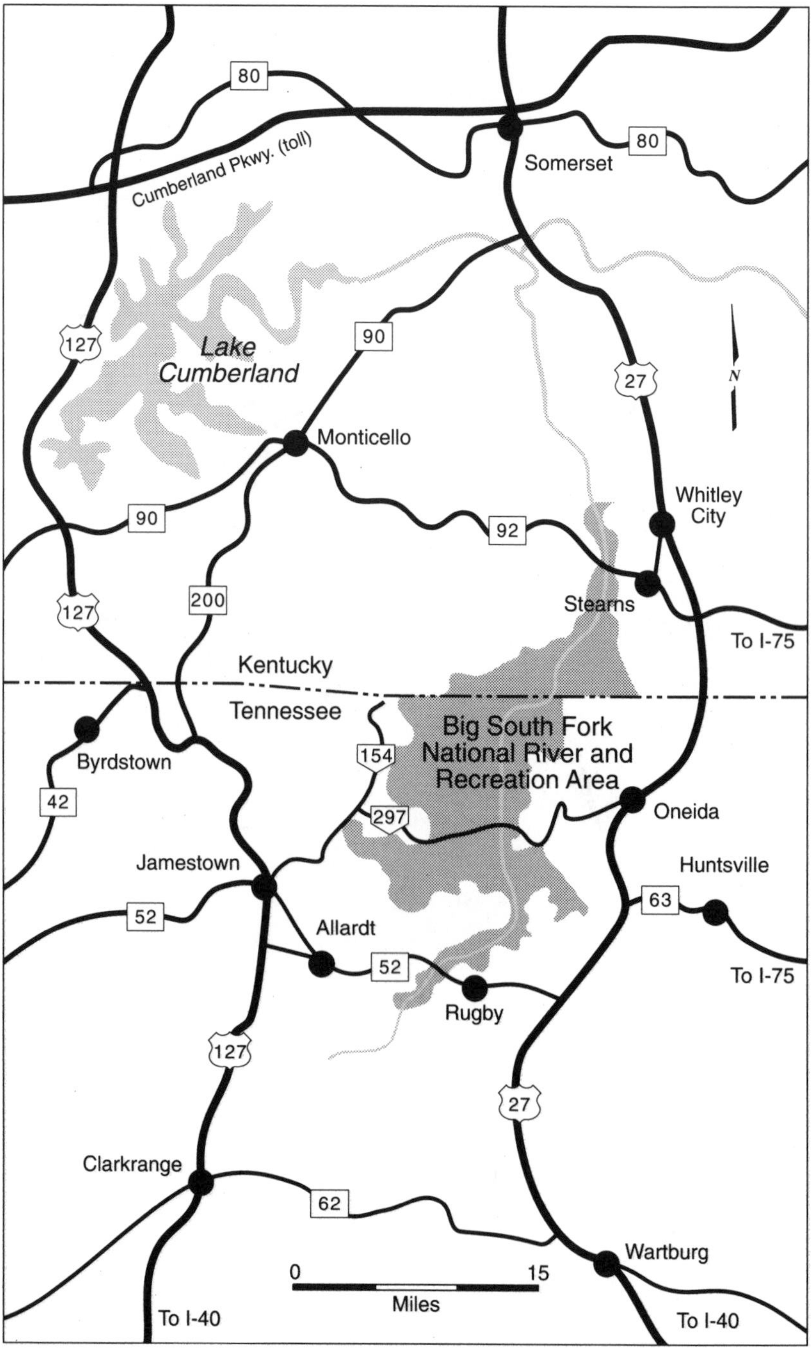

Map 16. Generalized location map of the Big South Fork National River and Recreation Area and Vicinity.

Two of the most plentiful natural resources were timber and coal. By the late 1800s the demand for the region's timber and coal required the construction of a railroad line. In the 1870s the Southern Railway built a line across the plateau, connecting Chattanooga, Tennessee, with Cincinnati, Ohio, to facilitate the hauling of timber and coal to population centers (Coleman and Smith 1993). The Oneida and Western (O&W), a logging and mining railroad constructed in the plateau area of northern Tennessee to connect the towns of Oneida and Jamestown with the Southern Railway line, prospered during the 1920s but declined rapidly during the Great Depression and was finally dismantled in 1953 (Coleman and Smith 1993). The O&W bridge over the Big South Fork of the Cumberland River is still standing, a landmark in the Big South Fork National River and Recreation Area.

Vast deposits of coal lay beneath the forested surface of the Cumberland Plateau. These coal deposits originated as accumulations of vegetative matter during the Pennsylvanian age, over 200 million years ago. Large swamps, containing forests of tree-like ferns and other plants, were the locations of the future coal beds. Because these ancient swamps were located near a coastal environment, they were often covered by sand deposits washed inland by coastal storms or meandering streams. These accumulations of sand are now the layers of sandstone that form the numerous bluffs, overhangs, and ledges of the Cumberland Plateau landscape.

From its beginnings near Rockwood in 1814, coal mining in Tennessee has developed into one of the most important mineral industries in the state (Floyd 1965: 35). During the first half of the twentieth century, coal mining boomed in the Cumberland Plateau region.

Coal is mined by two methods. In strip mining the surface soil and the top layers of rock are removed, exposing the coal beneath. In deep mining coal is extracted by excavating tunnels into a mountainside along a layer of coal. Numerous coal seams, varying in thickness from a few inches to over three feet, are found on the Cumberland Plateau, scattered throughout the Pennsylvanian rock layers. These deposits are bituminous coal. This soft coal, high in carbonaceous matter, has a moderate to high volatility (Floyd 1965: 35). Although coal mining has provided great economic benefits for some individuals and companies, it has taken a heavy toll on the health of the miners and the environment. Working in underground mines has proved hazardous for miners, not only from the immediate threats of cave-ins and explosions but also in the long-term results of breathing the coal dust, "black-lung" disease.

For the environment, strip mining in particular has been devastating. Stripping away vast amounts of vegetation, soil, and rock not only destroys habitats of plants and animals but causes pollution of streams.

Silt and clay eroded from strip-mine areas wash into the streams, filling the channels with sediment and contaminating the water with acid-mine drainage.

Several sites of former coal mines on the Cumberland Plateau are found within the borders of the Big South Fork National River and Recreation Area (Coleman and Smith 1993). Some of these are in northern Tennessee—the Zenith Coal Mine, the John Hawk Mine, and the Anderson Mine, for example. Just across the state line in Kentucky is the site of the Blue Heron coal mining camp, built by the Stearns Coal and Lumber Company.

The Blue Heron Visitor Complex, opened by the National Park Service in 1989, features a coal tipple (a building with conveyers to load coal onto railcars) dating from 1937. Frames of buildings typical of a mining camp have been constructed, and there are accompanying interpretive exhibits. The Stearns Scenic Railroad offers a 6-mile train ride, especially popular for viewing fall color in the wilderness of the Big South Fork.

Route details. The route begins along a relatively flat portion of the plateau. Pastures interplay with forests as the road winds along the tabletop plateau. Bedrock is most commonly seen along the creeks and the deeply incised river canyons, where the layers of sandstone and shale are exposed. Along the side of the road at No Business Creek (Log Mile 6.7) and Clear Creek (Log Mile 12.3) you can get especially good looks at the bedrock. The sandstone layers, generally gray to brownish-gray in color, vary in thickness from a few inches to over three feet. The shale, typically medium to dark gray, is very thin-layered, often breaking into small chips the size of a fingernail. The weathered shale is usually light brown and very clayey in texture.

Close to the city of Jamestown the route passes near two small towns of historic interest: Rugby and Allardt. Rugby is a Victorian village established as an experimental community in 1880 by the writer and social reformer Thomas Hughes. Seventeen of the original buildings still stand, beautifully restored (Fig. 102). The nearby village of Allardt, settled by German immigrants, dates from about the same period as Rugby. Several of the houses in Allardt have also been restored (including the Gernt House, which offers bed and breakfast).

Located just east of Allardt and 11 miles west of Rugby on S.R. 152 is Colditz Cove State Natural Area. Among its beauties is Northrup Falls, 60 feet high, which you see if you take the 1.5-mile loop hiking trail.

The route continues through Jamestown, where the parents of Samuel Langhorne Clemens (Mark Twain) once lived. Of special historic interest in the Jamestown area are a number of sites associated with Sgt. Alvin York, the great World War I hero.

Just north of Jamestown the route leaves U.S. 127 and takes you northeast along S.R. 154 to the natural scenic areas of Pickett State Park and the Big South Fork National River and Recreation Area. Now you are near the edge of the plateau, where streams have cut numerous connecting gorges and canyons, opening up the geology for all to see and enjoy (Figs. 103 and 104).

At Log Mile 44.7 the route turns right from S.R. 154 onto a gravel Park Service road going to the Twin Arches trailhead. This narrow gravel road mostly follows the flat topography except where the road turns down onto a finger of the plateau leading to the Twin Arches. Even at this point visible bedrock is scarce, but the topography of the land reflects the flat-lying character of the bedrock (Fig. 105). The route ends at a cul-de-sac where the foot trail to Twin Arches begins.

Fig. 102. Christ Church, Episcopal (built 1887), Rugby, Tennessee.

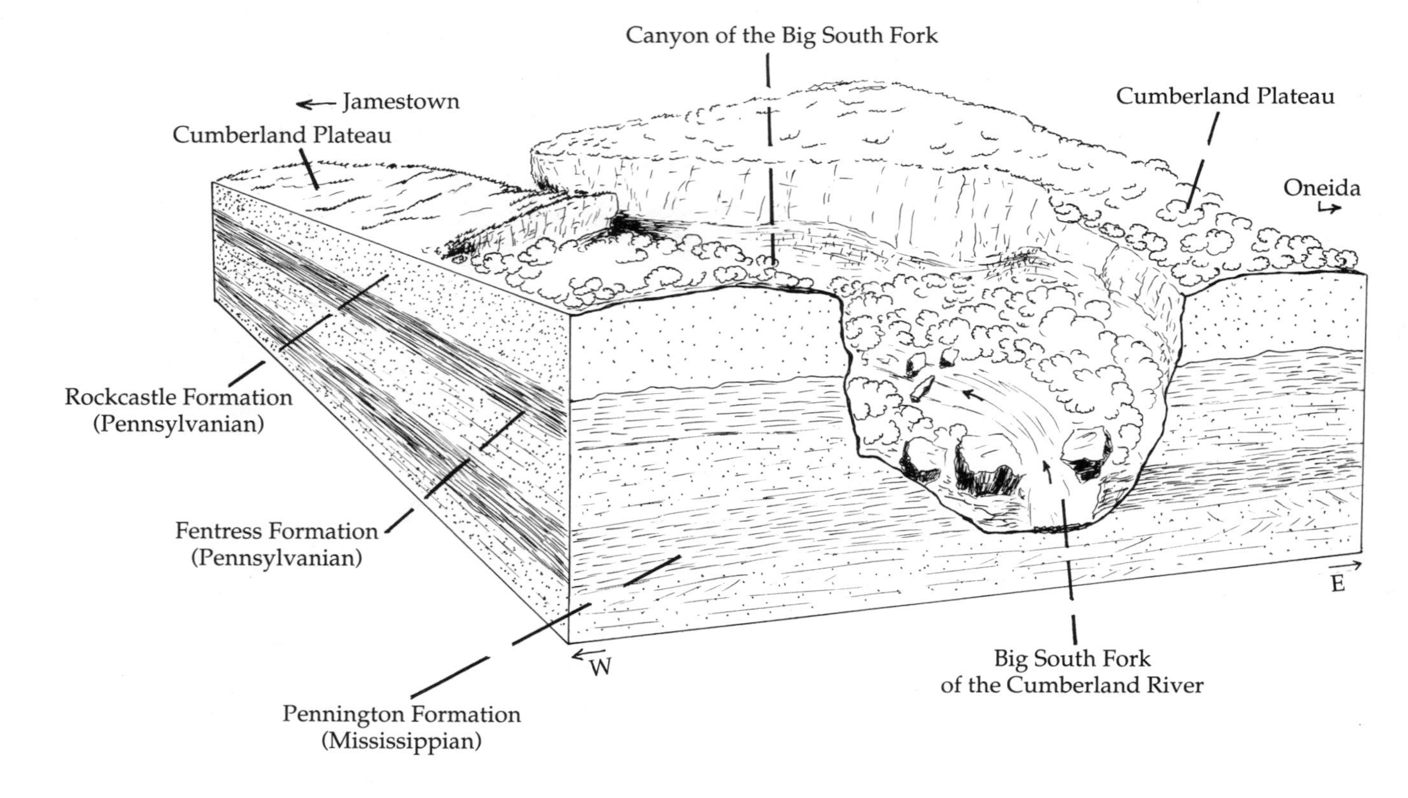

Fig. 103. Rock strata in the Big South Fork National Recreation Area consist of flat-lying Pennsylvanian-age strata as illustrated in this schematic drawing.

Fig. 104. Indian Rockhouse (Pickett State Park). Carved out of massive sandstone bedrock, this natural rock shelter is one of many in Pickett State Park and the Big South Fork National River and Recreation Area.

Fig. 105. Massive sandstone bluffs are a recurring feature along the rim of the plateau in the Big South Fork National River and Recreation Area.

Twin Arches Hiking Trail (optional)

Destination: Twin Arches
Location: Trail begins at end of Twin Arches Road, a gravel road in the Big South Fork National River and Recreation Area, located off S.R. 154
Start: Twin Arches Trailhead parking area, elevation 1,510 feet
End: Twin Arches, elevation 1,500 feet
Nature of the trail: Unpaved foot trail, moderately difficult; terrain ranges from fairly level to rolling, with several steep stairways
Distance: 1.4 miles round-trip
Special features: Spectacular sandstone rock arches, among the largest in the eastern United States; Pennsylvanian age sandstone and shale exposures; flat-lying sedimentary rocks; breathtaking views of the canyon

This hiking trail takes you to a most unforgettable natural landscape—twin rock arches. The 1.4 mile round-trip hike is well worth the effort. You begin the trail on relatively flat ground with mountain laurel and blueberry undergrowth. Quickly you encounter the plateau rim, where steep wooden stairs take you down over the face of the sandstone cliff. Many ferns and wildflowers grow along the moist, shady base of the sandstone rim rock. Notable geologic features of this trail include of the sandstone bedrock, with its unusual weathering characteristics, and massive exposures of the flat-lying Rockcastle Sandstone and Fentress Shale. This is a particularly lovely hike during October, when fall colors are at their peak.

Trail description. The trail begins as a moderately level walk through a mixed hardwood and pine forest. After about 0.1 mile the trail drops at a moderate grade to the top of a sandstone layer that forms the cap for the surrounding plateau tablelands (signs warn the visitor about the danger of falling from the edge of nearby cliffs).

At 0.2 miles the trail forks and a sign directs you to the left, toward the Twin Arches. The other fork will be your return trail from the Twin Arches. After turning left, the trail proceeds down the slope to a set of steep wooden steps that lead down over the face of the Rockcastle Sandstone Formation (Fig. 106). As you go down the steps, you can see the horizontal stratification of the rock, which indicates its sedimentary origin. Here the sandstone varies in color from light gray to a rusty orange, reflecting the presence of iron-bearing minerals such as pyrite and limonite.

Fig. 106. Along the trail to the Twin Arches, steep wooden staircases lead downward over sandstone bluffs.

From the base of the steps you can observe the nature of the sandstone bluff that the trail follows to the Twin Arches. The numerous sweeping diagonal lines you see in the sandstone are what geologists refer to as cross-bedding: the result of the rapid deposition of sediments (by oceans, streams, or wind), causing the headward movement of those sediments (usually sand) in the form of a ripple or wave—commonly found along ocean shorelines, river deltas, and sand dunes.

The trail continues to the right along the base of the sandstone bluff. Numerous wildflowers and ferns can be found along this section of the trail. At about 0.4 mile the trail crosses a spring flowing out from the base of the bluff. This spring is the result of a perched water table: groundwater seeping down through the porous sandstone reaches the top of an underlying shale formation. Because this shale, known as the Fentress Formation, is very tight and impermeable, the groundwater stops its downward movement at the top of the shale strata. There the groundwater begins a lateral movement along the shale and sandstone interface. Spring seeps occur where the groundwater reaches the surface along the base of the bluff. The shale strata can be identified by its very thin fissile layers and dark gray to blackish color.

The trail temporarily swings away from the bluff but rejoins the sandstone escarpment at 0.6 mile. The trail then swings left and brings you to the first arch (the North Arch) at 0.7 mile. The Twin Arches, the result of erosional processes, are the most spectacular geologic occurrence in the Big South Fork. Aligned next to each other, end on end, the Twin Arches form a rock structure so large that it is virtually impossible to view the entire natural bridge complex from one vantage point. The North Arch, the smaller of the two, has a span of 93 feet and a clearance of 51 feet (Fig. 107). The South Arch has a span of 135 feet and a clearance of 70 feet. Both arches are formed out of a Pennsylvanian age sandstone and conglomerate (with rounded pebbles) referred to by geologists as the Rockcastle Formation (Corgan and Parks 1979: 75).

At the North Arch the rusty orange-red color of the rock is due to iron staining from the chemical breakdown of the iron sulfide minerals in the rock mass. As the minerals in the rock are exposed to water and air, oxidation takes place, leaving a rusty residue within the rock along the surface.

The most notable features of the South Arch (Fig. 108) are the joints (cracks) within the rock mass. Weathering along these cracks has removed the material holding the sand grains together. The result is that the cracks widen as the rock continues to weather. Some of these

Fig. 107. The North Arch. Sandstone strata of the Rockcastle Formation have been weathered to form the Twin Arches.

cracks have been enlarged enough to form tunnels, such as the one along the east end of the South Arch. In places these joints can widen enough to create an arch structure in the rock strata.

The Twin Arches themselves are the result of weathering processes. Erosion of the plateau area by rain, sleet, snow, freeze and thaw cycles, and heat has entrenched deep gullies and canyons in the bedrock and enlarged cracks within the strata. Headward erosion of the gullies and canyons has encroached upon the plateau caprock, exposing the sandstone strata along the rim of the plateau.

Because the groundwater flows down through the sandstone and stops at the top of the impermeable shale strata (just beneath the caprock), the groundwater saturates the base of the sandstone, dissolving the cementing material in the sandstone. As the base of the sandstone layer is eroded at a faster rate than the top layer of sandstone, an overhanging shelf is created. This feature can be seen at the numerous rock shelters in the area, such as Indian Rockhouse and Hazard Cave.

As this type of headward erosion of the gullies and ravines continues toward the top of the ridge, the result will eventually be a breach of the sandstone caprock. In some instances continued enlargement of the overhanging rock shelters on either side of a ridge will eventually result in the intersection of the rockhouses, forming an arch out of the

Fig. 108. The South Arch. The massive 135-foot span of the arch dwarfs two hikers.

overlying sandstone cap. Such was probably the case at Twin Arches, where the unusual result was two arches being formed side by side.

To return to the parking area you can retrace your route to the trailhead. If you are adventurous, and not prone to acrophobia, you might consider a return route over the top of the arches. For this route, you must climb the stairs located between the North and South Arch, which will bring you to the top of the arches. The sandstone caprock, about 25 feet wide at this point, is exposed at the surface and the trail leads right over the top of the North Arch. *Caution:* Because the edges of the rock ledge are sudden and are not protected with handrails, you must be extremely careful here. This route is not recommended if you have small children along.

Leaving the top of the North Arch, you quickly encounter a series of steep steps taking you up to the top of a small summit along the ridge. From the top of the stairs you can see the surrounding plateau rim with numerous large sandstone bluffs. A view from this point during the fall color season will be very rewarding. (Reindeer moss, a grayish lichen, can be found all along this section of the trail.) The trail continues along the ridge and connects back to the main trail just above the first set of wooden stairs. The parking area and trailhead are 0.2 mile from this junction.

For further exploration of the beautiful, geologically fascinating Big South Fork National River and Recreation Area and its over 200 miles of hiking trails, you may wish to consult Brenda D. Coleman and Jo Anna Smith, *Hiking the Big South Fork,* second edition (Knoxville: Univ. of Tennessee Press, 1993), or Russ Manning and Sondra Jamieson, *The Best of the Big South Fork* (Norris, Tenn.: Laurel Place, 1989).

Additional information (food, lodging, schedules of events):

Historic Rugby
P.O. Box 8
Rugby, TN 37733
(615) 628-2441

Big South Fork National River and Recreational Area
National Park Service
P.O. Drawer 630
Oneida, TN 37841
(615) 879-3625

Pickett State Park
P.O. Box 174
Jamestown, TN 38556
(615) 879-5821

Fentress County Chamber of Commerce
P.O. Box 496
Jamestown, TN. 38556
(615) 879-9948

4. Fall Creek Falls State Park

Destination: Fall Creek Falls, Fall Creek Falls State Park
Route: U.S. 127, S.R. 30, S.R. 284, and Park Road; begins at I-40, Exit 317 (Map 17)
Cities and towns: Crossville, Pikeville
Trip length: 45.3 miles to north park entrance (one-way); another 5.4 miles to trailhead via park roads (total 50.7 miles one-way)
Nature of roads: Two-lane, paved highways, mostly straight with some curvy and hilly sections; very scenic drive
Beginning elevation: 1,778 feet
Ending elevation: Fall Creek Falls Overlook: ± 1,640 feet
Special features: Plateau topography; highest waterfall in Eastern United States; Sequatchie Valley; Pennsylvanian age sandstone and shale
Hiking trail (optional): Base of Fall Creek Falls
U.S.G.S. 7 1/2 minute quadrangle maps: Sampson 103 NE, Spencer 103 NW
For reference: Plateau, Pennsylvanian age, Sewanee and Warren Point formations, Sequatchie Valley

Directions. From I-40 take Exit 317 onto U.S. 127. Go south for 31.1 miles, passing through Crossville. Turn right onto S.R. 30 (west) and follow it for approximately 12.2 miles. Then turn left onto S.R. 284 (west) and follow it approximately 2 miles to the Fall Creeks Fall State Park entrance. From this north entrance to the park, travel 1.3 miles to a stop sign (to the right is the nature center, to the left on S.R. 284 is Park Headquarters). Continue straight for 0.9 miles, then veer right at the camper check-in station. After bearing right, continue on the Park Road for 1.6 miles, passing the lake and crossing the dam. At 1.6 miles turn right

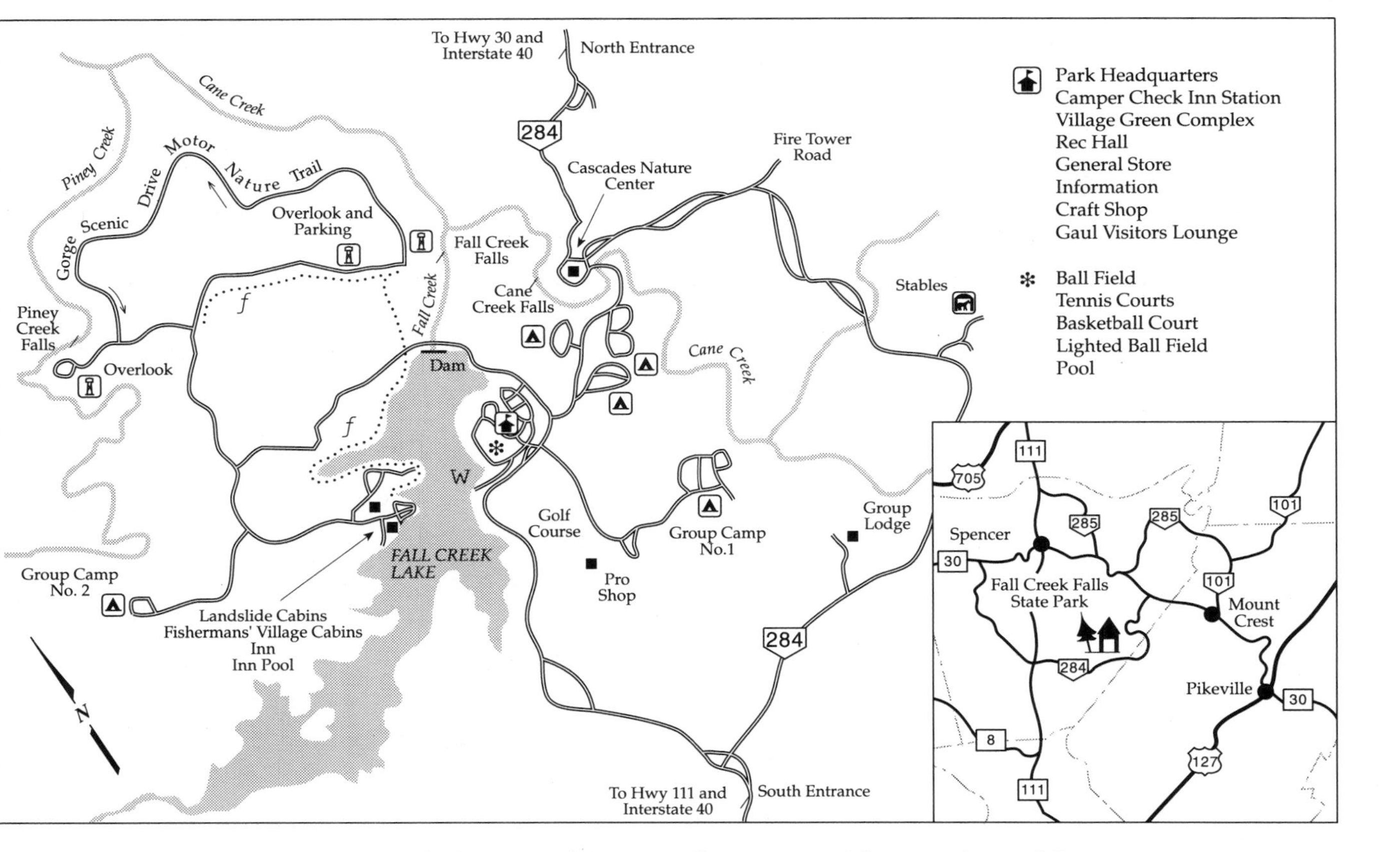

Map 17. Fall Creek Falls State Resort Park. Courtesy of Tennessee Department of Conservation and Environment, State Parks Division.

onto the self-guiding scenic drive and continue for 1.6 miles to the parking area for Fall Creek Falls Overlook and the trailhead to the base of the falls.

Route details. Along this route you will see the flat to rolling landscape along the top of the Cumberland Plateau, enjoy the scenic beauty of Sequatchie Valley, travel through the historic area of Homestead, and view the bluff-lined gorges and waterfalls of Fall Creek Falls State Park.

The route begins along a relatively flat portion of the plateau. After you pass through the Crossville area, the plateau becomes somewhat more hilly, crossing several creeks (Basses Creek at Log Mile 11.8 and Daddy's Creek at Log Mile 15.0). Near Log Mile 7.6 on U.S. 127 is a community called Homestead. The many stone houses found here are the results of a homesteading project of the Roosevelt era following the Great Depression. The Homesteads Tower, visible from the highway, was built in 1937 and 1938 to house the administrative offices of the Cumberland Homesteads Project.

The stone used to construct the houses and tower are cut from the thin-bedded Crab Orchard Sandstone Formation. Characteristic of this sandstone are the iron staining patterns developed along the sand layers. This building stone has been extensively quarried and marketed throughout the Southeast. A few sandstone outcrops are noticeable before the route takes you down into Sequatchie Valley near Log Mile 17.2 on U.S. 127 in Cumberland County. The long, narrow floor of Sequatchie Valley, which lies about 800 to 1,000 feet below the plateau rim, consists of residual clay soil and limestone strata.

Sequatchie Valley (Fig. 109) is a unique feature in the plateau landscape. The bedrock in the valley, unlike the flat-lying rock on the top of the plateau, has been folded and faulted into a long, narrow upward bend that runs from near the Crossville area in Tennessee southwest into northern Alabama. This upward fold, called an anticline, has been breached by weathering along the crest of the fold, exposing the more soluble limestone beneath the surface of the plateau to weathering forces as well (Fig. 110). Over many millions of years, erosion has produced a long, narrow valley parallel to the long axis of the fold; this valley, too, extends southwestward into northern Alabama.

The headward development of Sequatchie Valley is carried out by the solution activity of the groundwater on the underlying limestone. As the limestone is dissolved through chemical solution activity by the groundwater, sinkholes and caves form at the surface. Some of the

Fig. 109. Sequatchie Valley is underlain by limestone strata that weather to provide fertile soils, which are commonly farmed.

sinkholes continue to enlarge and interconnect, forming a large sinkhole feature called a uvala.

Grassy Cove, located just southeast of Homestead on S.R. 68 is a large uvala that contains several cave systems. If you would like to visit Grassy Cove, turn off U.S. 127 at Log Mile 7.6 (Homestead) and follow S.R. 68 for about 10 miles. The cove, which is very large, looks like a valley as you drive through it. Surface water draining into the sinkholes in Grassy Cove flows through the subterranean cave systems and reemerges in Sequatchie Valley to form the Sequatchie River.

The drive down the valley passes rolling pastures bounded by the rock bluffs of the valley walls. The route climbs the western side of the valley wall just west of Pikeville and continues along several large sandstone bluff overhangs (near Log Mile 33.9 on S.R. 30; Fig. 111). At the plateau rim the route proceeds westward across the plateau proper, and again the landscape is flat to rolling. You do not see massive exposures of bedrock again until you enter the gorge areas of Fall Creek Falls. Fall Creek Falls State Park is in two counties, Bledsoe and Van Buren. The gorge and waterfalls are in the Van Buren portion of the park.

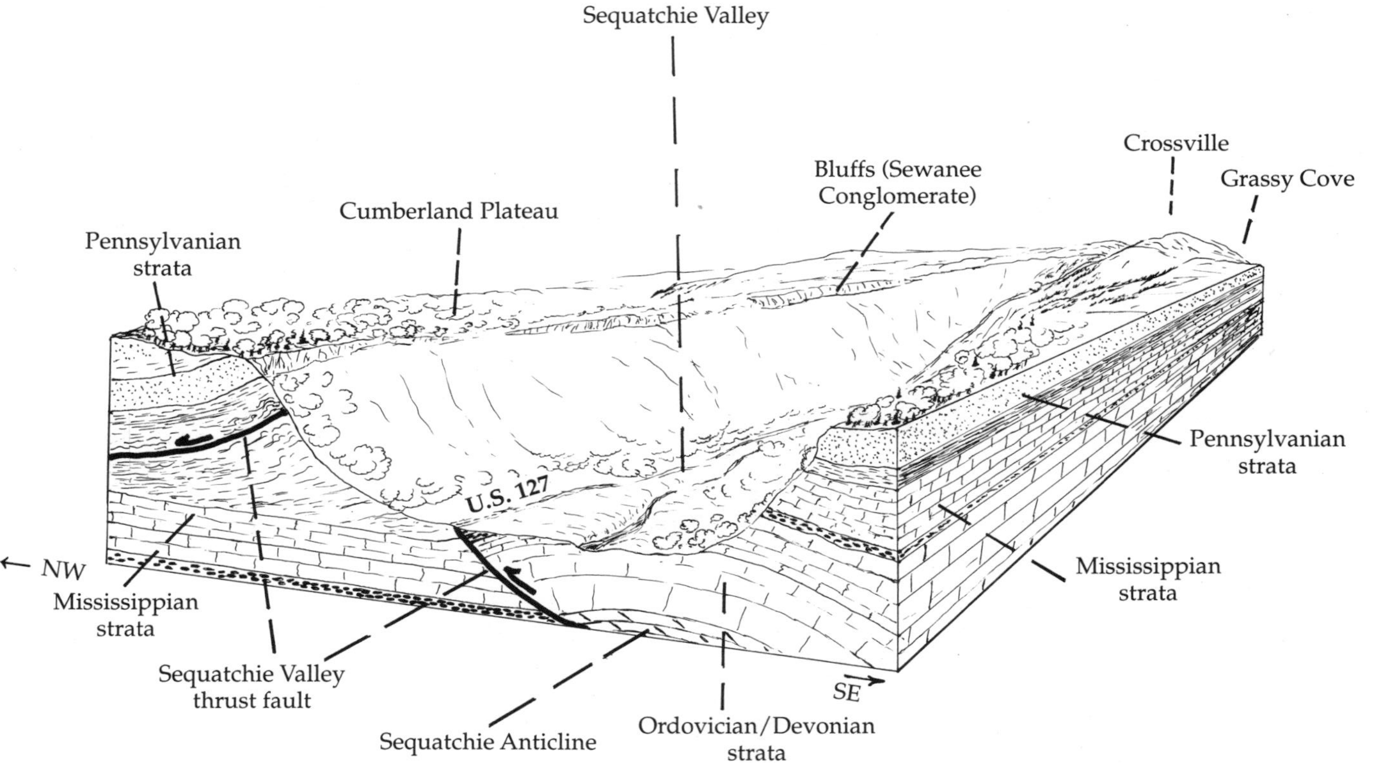

Fig. 110. Sequatchie Valley formed from the weathering of rock strata that have been folded in an upward arch (anticline), as illustrated in this diagram. Note the valley formed along the crest of the anticlinal structure.

Fig. 111. Along the upper rim of Sequatchie Valley you may notice massive outcrops of Pennsylvanian-age sandstone, which form the caprock of the Cumberland Plateau (Log Mile 33.9).

At the north entrance to Fall Creek Falls State Park the elevation is 1,819 feet and the landscape seems relatively flat. But after about 1 mile the road quickly drops in elevation to 1,611, entering the upper drainage of Cane Creek. Here at the intersection of two park roads is the nature center, which contains an excellent geologic exhibit as well as displays on other facets of the park. Park headquarters are to the left, along S.R. 284.

Several hiking trails originate at the rear of the nature center. These include a short 0.4 mile (round-trip) hike to the foot of the Cane Creek Falls (very difficult) and a 2-mile round-trip hike to the base of Fall Creek Falls via the Fall Creek Falls Overlook (easy to moderately difficult).

The road straight ahead (not going left to park headquarters or right to the nature center) takes you by the camping areas and past Fall Creek Falls Lake and Dam. About 0.5 mile past the dam, the road forks: the right fork takes you on the scenic auto drive leading to the Fall Creek Falls Overlook and trailhead, the left fork to the inn, restaurant and cabins.

Here in the park you are near the edge of the plateau, where Cane Creek, Piney Creek, and Fall Creek have cut deep canyons into the landscape. It is over the rim of the plateau that Fall Creek Falls, Piney Creek Falls, and Cane Creek Falls are developed. These natural gorges provide an excellent opportunity to view not only the geology but also various kinds of vegetation and wildlife. Thousands of years of erosion have resulted in those beautiful natural areas for all to enjoy.

A similar gorge, known as Savage Gulf, is found about 25 miles south of Fall Creek Falls. ("Gulf" refers to the open area of the gorge.) Now designated a State Natural Area, Savage Gulf contains several waterfalls and limestone caves.

Fall Creek Falls Hiking Trail (optional)

Destination: Base of Fall Creek Falls
Location: Trail begins at Fall Creek Falls Overlook, located along Fall Creek Falls self-guiding scenic drive
Start: Top of Fall Creek Falls, elevation ± 1,580 feet
End: Plunge pool of Fall Creek Falls, elevation ± 1,324 feet
Nature of the trail: Unpaved foot trail, rocky and moderately difficult; several switchbacks and steps
Distance: About 1 mile round-trip from overlook parking area
Special features: Spectacular waterfall, highest in Eastern United States at 256 feet; massive exposures of Pennsylvanian age sandstone; breathtaking views of Fall Creek and Cane Creek Gorge

This moderately difficult but rewarding hiking trail takes you down into the canyon environment of Fall Creek Gorge and to the base of Fall Creek Falls (Fig. 112). The approximately one mile round-trip hike leads past a stand of virgin hemlocks and tulip poplars, moss-covered boulders (some house-sized) and sheer cliffs. You begin the trail atop the gorge at the Fall Creek Falls Overlook and end at the base of the waterfalls. Exposures of two Pennsylvanian age formations are found along the trail: Sewanee Conglomerate and Warren Point Sandstone.

Trail description. The hike to the base of the falls begins with a breathtaking view of the falls themselves, the highest in the eastern United States. At the overlook you can see Fall Creek as it gently glides along the rusty colored sandstone and then plunges over the edge of the canyon. A drop of 256 feet terminates in a plunge pool lined with dark, rusty colored breakdown from the overhanging sandstone.

The view to the left of the overlook is dramatic indeed: you can see both the Cane Creek Gorge and a rocky pinnacle that overlooks Fall

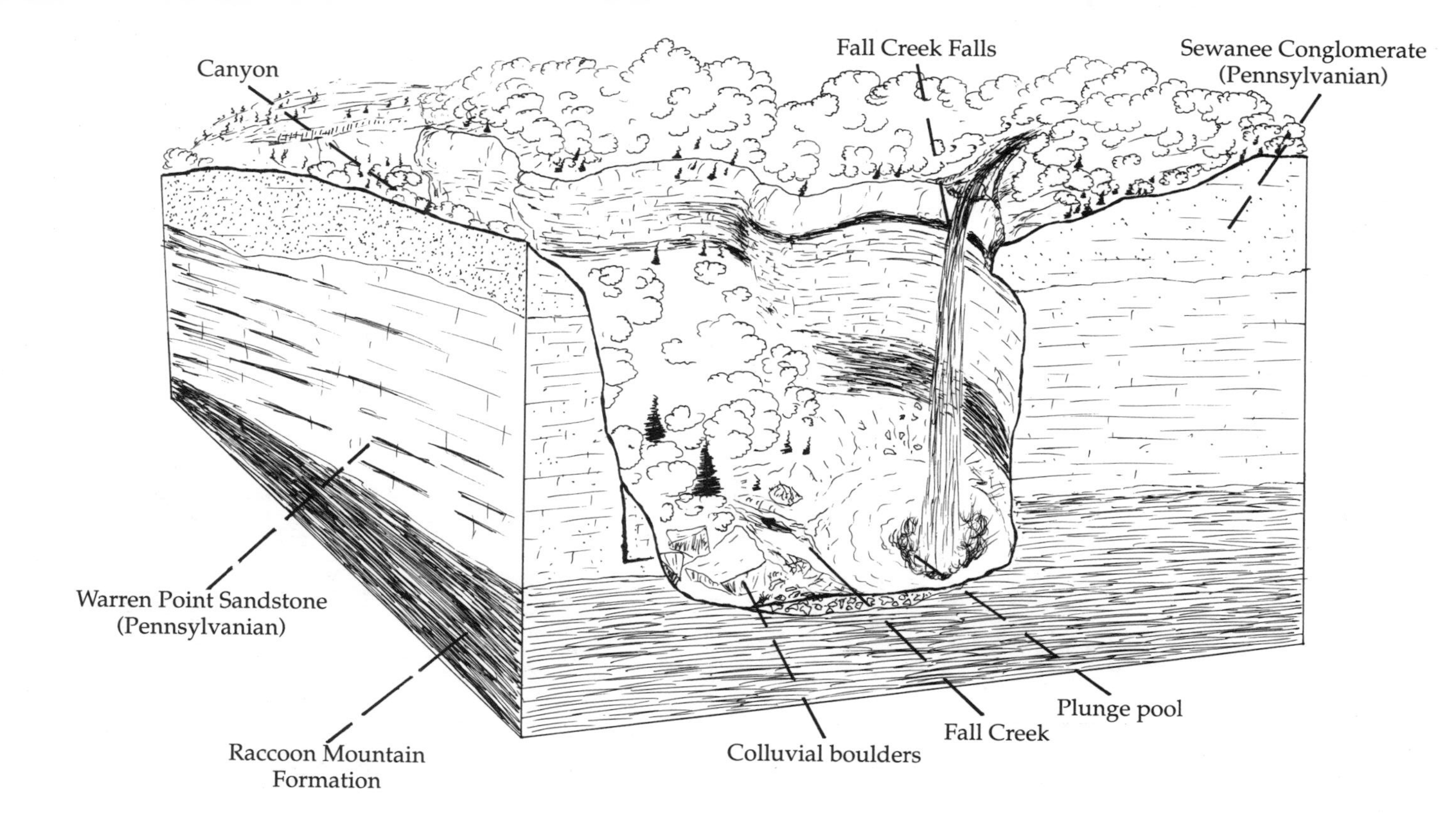

Fig. 112. Fall Creek Falls is formed where Fall Creek flows over Pennsylvanian-age strata into a canyon carved by Cane Creek and Fall Creek.

Creek. Bluffs, yellowish-brown to gray in color, line the canyon through which Cane Creek flows. From the base of the falls, Fall Creek flows approximately 0.3 mile before joining Cane Creek.

From the overlook, the trail and the base of the falls are to the left. The trail is at first a gentle slope. After about 50 yards, the trail comes to a second overlook, where you can gain a better perspective of the Cane Creek Gorge.

At this point the trail steepens considerably as you descend the canyon walls. Because the trail becomes very rocky and narrow, you must give careful attention to where you place your feet.

About 200 yards from the beginning, the trail comes to the first of several switchbacks (places where the trail zigzags to make the descent and climb back out less difficult). At this point you can see large sandstone bluffs to the left of the trail. Notice that the rock strata appear to be stacked—the result of the thin-layered sandstone beds being broken through natural processes to form joints.

As you make the first switchback, look at the numerous angular blocks of grayish sandstone lying on the surface of the ground. This is colluvium: gravity-deposited material, such as talus. When the blocks of rock are spread out laterally, in a sheet, the deposit is referred to as a block field. Such is the case here. (When the colluvial blocks are deposited in a long, linear fashion—up and down a slope—the deposit is referred to as a block stream.)

About 0.25 mile along the trail, you encounter a very large rock face. This rock exposure consists mainly of sandstone and conglomeratic sandstone belonging to the Sewanee Conglomerate and Warren Point Sandstone formations. The uppermost layers, a gray to brown sandstone containing numerous characteristic pea- to marble-sized white quartz pebbles, are the Sewanee Conglomerate. The layers below, brown to gray in color and fine to medium in grain, are Warren Point Sandstone, a formation with interbeds of gray shale, some of which contain coal beds.

Along this rock face you will notice a very large vertical crack (joint) extending over 100 feet up the rock exposure. This is a good example of how the forces of erosion and weathering attack such a formidable rock mass. Joints in the rock mass are enlarged through the action of moving water and, during the winter season, "ice jacking": water caught in the cracks of surface exposures of rock freezes and consequently expands, wedging apart the two sides of the rock joint. When a particular joint is attacked by these erosional forces, a large rock mass can be broken, moved, and eventually toppled. During the summer months cool air will blow out of the base of the crack—a natural air conditioner.

As you follow the steps down along the rock face to the base of the rock bluff, you quickly encounter an area of overhanging rock faces (Fig. 113). These overhangs are formed by erosion of the weaker shale and shaly siltstone layers, part of the Raccoon Mountain Formation that underlies the massive Warren Point Sandstone Formation in this area.

Fig. 113. The trail down to the base of Fall Creek Falls leads past massive exposures of Pennsylvanian-age strata.

At a point along the overhang bluff about 0.3 mile from the beginning of the trail, you can see another large crack in the bluff. Here, too, cool air issues from the fissure during the summer months. About 100 yards farther and you come to the plunge pool of Fall Creek Falls. Before you is 256 feet of falling water, which has carved out an almost circular plunge pool beneath it (Fig. 114). The falls have developed over thick, hard, well cemented sandstone layers of the Sewanee Conglomerate and the underlying Warren Point Sandstone (Fig. 115). The plunge pool has developed in the weaker shale and siltstone strata within the lower part of the Warren Point Formation.

Particularly noticeable is the rusty color staining the rock in the plunge pool area. This color results when a mineral known as pyrite (fool's gold) is broken down and oxidized. In some places, especially in the creekbed above the falls, a precipitate of the chemical breakdown of iron sulfide develops. This precipitate occurs as a rusty-orange deposit of iron hydroxide (known locally by coal miners as "yellow boy").

Though now well inland, the rocks exposed in the canyon walls tell a story of coastlines, sandy islands, and swamps. These rocks were formed along an ancient coastline some 250 to 300 million years ago. Along

Fig. 114. Around the plunge pool of Fall Creek Falls you can see the iron-staining effects from the weathering of the mineral pyrite.

Fig. 115. Fall Creek Falls, 256 feet in height, is the highest waterfall east of the Mississippi River.

this ancient coastline numerous barrier islands (long, sandy islands parallel to the coast) sheltered swampy lagoon areas landward of the barriers (Ferm, Milici, and Eason 1972: 4–6). There were extensive tidal flats interwoven between the lagoons and the shore. When storms battered the islands and stripped the vegetation, sand and vegetation debris were deposited into the lagoons. There were also numerous creek channels that drained mainland water out to the sea between the barrier islands.

As time passed, the barrier islands left deposits of sand as they migrated along the coast. These sand deposits are now known as sandstone; some are conglomeratic, containing rounded, white quartz pebbles. The tidal flats and lagoon deposits became shale and siltstone; where extensive vegetation accumulated, a coal seam developed. Plant fossils of some of the vegetation can be found in some of the sandstone and shale layers. Two of the more common plant fossils are *lepidodendron* (fern bark) and *calamites* (a plant similar to a bulrush). Today the forces of water are still at work, weathering down these masses of sandstone and carrying the sand back to the ocean, possibly to become another barrier island.

To return to the trailhead, retrace the trail back to the top. As you rest going up the steep incline, notice the large hemlock and tulip poplar trees, which represent virgin forest in this area. Many ferns and wildflowers can also be seen along the trail.

For further exploration of the beautiful Fall Creek Falls area, you may wish to consult Evan Means, *Tennessee Trails* (Chester, Conn.: The Globe Pequot Press, 1989), or Russ Manning and Sondra Jamieson, *The South Cumberland and Fall Creek Falls* (Norris, Tenn.: Laurel Place, 1990).

Additional information (food, lodging, schedules of events):

Fall Creek Falls State Park
Route 3
Pikeville, TN 37367
(615) 881-3297

Cumberland Homesteads Museum
Route 12, Box 96-C
Crossville, TN 38555
(615) 456-9663

5. Burgess Falls State Natural Area

Destination: Burgess Falls State Natural Area
Route: S.R. 135 (Map 18)
Cities and towns: Cookeville
Trip length: 8.7 miles one-way; to return to I-40, retrace the route
Nature of the roads: two-lane, paved highway, curvy and hilly
Beginning elevation: ± 1,090 feet
Ending elevation: ± 875 feet
Special features: Eastern Highland Rim landscape, gorge of Falling Water River, Devonian and Mississippian age rock strata, flat-lying sedimentary rocks
Hiking trail (optional): Burgess Falls
U.S.G.S. 7 1/2 minute quadrangle maps: Burgess Falls 326 SE
For reference: Eastern Highland Rim, Devonian and Mississippian age, waterfalls

Directions. From I-40 take Exit 286 onto S.R. 135. Go south on S.R. 135, leaving the Cookeville city limits and entering the rolling countryside. Follow S.R. 135 for 8.5 miles to the White County line; turn right onto the paved road and follow it for 0.25 mile to the entrance to Burgess Falls State Natural Area. The parking area for the Burgess Falls trail is 0.1 mile inside the entrance.

Route details. As you travel along this route, you will notice a distinct absence of rock exposures because the area is underlain by Mississippian age limestone, which weathers to produce a distinctive, reddish-orange, residual clay soil. Rock exposures are generally found along creek and river channels and the Highland Rim escarpment.

The route begins as you turn onto S.R. 135 (Log Mile 0.0) and leads south, quickly leaving the commercial areas of Cookeville. In the landscape you will notice the rolling hills and the reddish clay soil typical of the Eastern Highland Rim.

These soils are derived from the chemical breakdown of the Mississippian age limestone: St. Louis, Warsaw, and Fort Payne formations. These three rock formations are all carbonates in composition; the differences between them are in the associated minerals and the type of carbonate sedimentation (that is, whether the origin of the sediment was a reef or carbonate mud, for example). As carbonates, all three of these

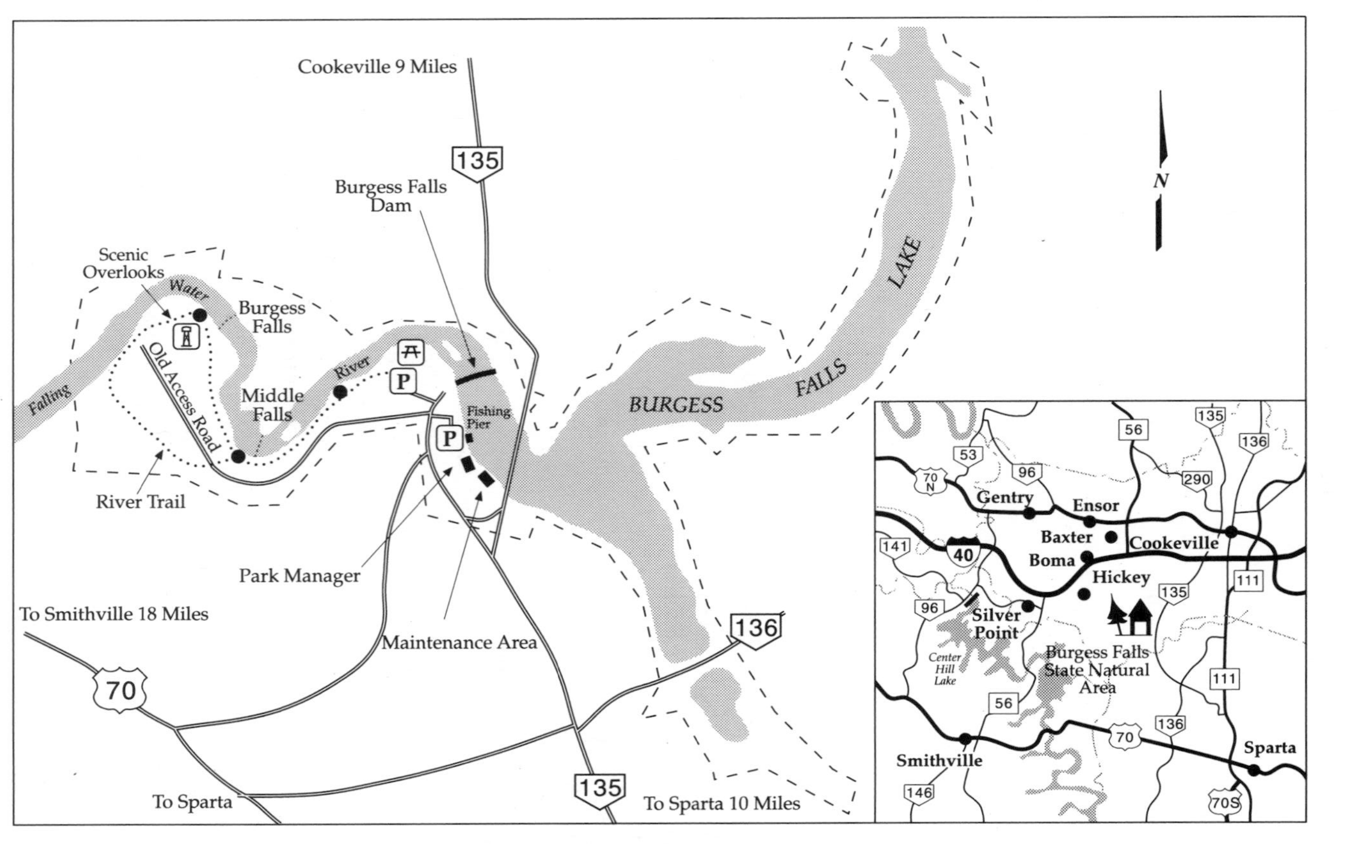

Map 18. Burgess Falls State Natural Area. Courtesy of Tennessee Department of Conservation and Environment, State Parks Division.

limestone formations are subject to the solution action of infiltrating acidic groundwater. Eventually the chemical reactions produce clay soil from the limestone. This soil is stained reddish-orange because of the presence of iron (usually small amounts, less than 1 percent in most cases).

Near Log Mile 4.9 there is a sharp curve to the left; follow S.R. 135 to the left. At Log Mile 8.3, S.R. 135 crosses Falling Water River, where Burgess Falls Dam has impounded the river to form Burgess Falls Lake. At the end of the bridge, you approach the White County line, where you turn right onto a paved road. Follow this road for 0.25 mile to the entrance to Burgess Falls State Natural Area.

After entering the park, follow the park road for 0.1 mile to the parking area for the Burgess Falls trail. This route brings you to the edge of the Eastern Highland Rim, where the Falling Water River has cut a gorge down through the Highland Rim escarpment. Exposed along the gorge walls and floor are flat-lying sedimentary rocks of Ordovician, Devonian, and Mississippian age. Several impressive waterfalls have formed over prominent outcrops of the rock strata. Burgess Falls, the highest of the waterfalls, drops 130 feet down into the backwater of Center Hill Lake (Fig. 116).

In 1793 the United States government deeded the land on which the falls are located to Tom Burgess as partial payment for his services in the Revolutionary War, services that included the building of a road across the Cumberland Mountains for the Federal Government. Burgess, of Rowan County in North Carolina, moved to Tennessee in 1814 (Burgess 1977: 3).

This area along Falling Water River became very important in the development of the region. A gristmill and sawmill were established in this area of the river in the early settlement days of the 1800s (Burgess 1977: 22). In the early 1920s a dam was constructed by the City of Cookeville across the Falling Water River to provide electricity, but in 1928 heavy rains caused the dam to break, and the powerhouse was destroyed. A new, stronger dam and a new powerhouse were constructed within a few years. Although this later dam still exists, it is no longer used to produce electricity; from 1944 on TVA's massive new dams and powerhouses have provided hydroelectric power for the area (Burgess 1977: 24–25).

Burgess Falls was designated a State Natural Area in 1980 to preserve its unique scenic, biologic, and geologic features for the enjoyment of everyone. The area is managed by the Tennessee Department of Environment and Conservation.

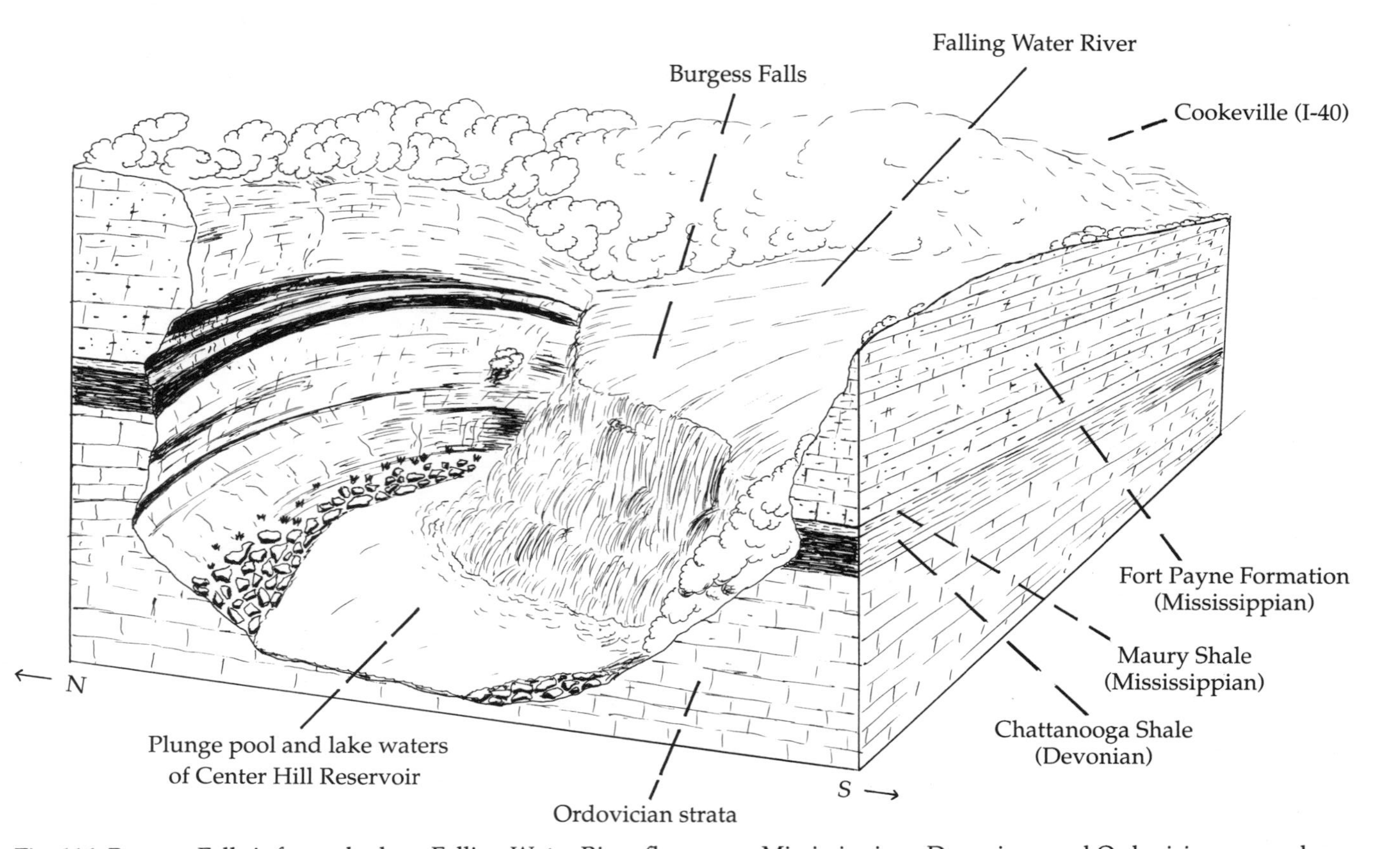

Fig. 116. Burgess Falls is formed where Falling Water River flows over Mississippian-, Devonian-, and Ordovician-age rocks along the edge of the Eastern Highland Rim.

Burgess Falls Hiking Trail (optional)

Destination: Burgess Falls
Location: Trail begins near the northwestern corner of the lower parking lot, close to the picnic pavilion
Start: Burgess Falls parking area, elevation ± 875 feet
End: Top of Burgess Falls, elevation ± 740 feet
Nature of the trail: Unpaved foot trail, easy to moderate difficulty; numerous steps; some steep sections
Distance: 1.5 miles, round-trip
Special features: Spectacular waterfalls, river gorge; Eastern Highland Rim escarpment; Ordovician, Devonian, and Mississippian age rocks; numerous wildflowers (spring)

The hiking trail takes you along the side and rim of the scenic gorge made by the erosive action of Falling Water River. The 1.5-mile hike is particularly rewarding because of the beauty of the waterfall. You quickly enter the gorge and then climb back up to the canyon rim, where you have fine overall views of the several waterfalls and cascades along the river. Along the canyon walls, especially as you go down to the base of Burgess Falls, you can see some excellent examples of outcrops of undisturbed flat-lying sedimentary rock strata of several ages.

Trail description. The trail begins by gradually descending the side of the canyon. There are several steps along this section of trail.

After about 75 yards you come to Little Falls Overlook. Little Falls is formed over a small ledge of silicious Mississippian limestone. Along the trail you can easily observe the limestone because there are numerous overhangs (Fig. 117). The dark gray to grayish-white nodules, golf ball–sized or fist-sized, within the limestone layers are deposits of the mineral quartz in the form of flint. It is the silica from these nodular deposits of quartz that makes the rock somewhat resistant to erosional processes; this rock forms the top layers of the hills around Burgess Falls.

You will begin to notice the presence of many hemlock, cedar, and mixed deciduous trees along both sides of the trail. After 0.38 mile the trail makes a steep climb, up many steps, back onto the canyon rim. At the 0.5 mile point you come to the dramatic overlook of Middle Falls. The view down into the canyon, over 150 feet deep at this point,

Fig. 117. The trail to Burgess Falls leads past numerous exposures of horizontally bedded Mississippian-age strata.

provides a good opportunity to see the full 200-foot width of the 50-foot-high Middle Falls (Fig. 118). In addition, you can easily observe the flat-lying, sedimentary character of the limestone.

Continuing on the trail toward Burgess Falls, you enter a mature forest with large hemlock and American beech trees (Fig. 119). The trees with the tall, straight, gray trunks and slightly square leaves are tulip poplar, the Tennessee state tree. Fall color at Burgess Falls State Natural Area is exceptionally good.

Near the 0.75 mile mark you come upon the breathtaking overlook for the 130-foot-high Burgess Falls (Fig. 120). Here you can also get a look at the overall makeup of the canyon's geology. From the top of the canyon to the bottom, you can see the following strata:

Mississippian limestone (Fort Payne Formation)—gray silicious and calcareous layers containing numerous rounded masses of chert and flint

Mississippian shale (Maury Shale)—light green to greenish-gray in color, 18 inches thick, thinly laminated, and clayey in texture

Devonian shale (Chattanooga Shale)—black, 20 feet thick, thinly laminated, hard, and platy

Ordovician limestone—gray, medium-bedded, calcareous layers.

Fig. 118. Middle Falls, as seen from the trail overlook, is formed where Falling Water River flows over resistant beds of the Mississippian-age Fort Payne Formation.

Fig. 119. The Burgess Falls trail passes through a scenic section of forest composed of tulip poplar, oak, and hemlock.

Fig. 120. Burgess Falls as viewed from the trail overlook is 130 feet in height.

Burgess Falls flows over the bottom few feet of the silicious Fort Payne Formation and cascades down over the Maury and Chattanooga shales as well as some of the Ordovician limestone that makes up the lower portion of the falls. The silicious limestone beds, which are resistant to erosion, provide the caprock, over which the water flows, forming the falls. The Maury and Chattanooga shales erode much faster than the silicious Fort Payne Formation, causing the Fort Payne strata continually to overhang the weaker shale strata.

Continuing down into the canyon from the overlook, the trail goes about 75 yards to the river level above the falls. *Caution:* This area is not protected by fences or railings. Be careful, especially if you have children along. Walking out to the edge of the falls is dangerous and is not recommended.

If you are adventurous, you may wish to follow the set of narrow steps and footpath that lead down the canyon to the bottom of the falls and to the edge of Center Hill Lake. Along this path you can get a close look at the greenish Maury Shale and the black, rusty-stained, fissile Chattanooga Shale. *Caution:* The steep, metal steps may be slippery, especially in the rain or when snow or ice is present.

The location of Burgess Falls will continue to move upstream as the erosive powers of the river waters continue to weather away the rock. In the not too distant future (geologically speaking), perhaps several thousand years, Burgess Falls may take on a new look or shape as joints in the bedrock widen and pieces of rock break away from the parent strata. Perhaps it will widen out to the full width of the canyon, or maybe become a free-fall waterfall. Whatever the result, the falls will no doubt continue to be a place of scenic beauty.

Additional information (food, lodging, schedules of events):

Park Manager
Burgess Falls Natural Area
Route 6, Box 380
Sparta, TN 38583
(615) 432-5312 or (615) 761-3338

Cookeville Area–Putnam County Chamber of Commerce
302 South Jefferson Avenue
Cookeville, TN 38501
(615) 526-2211

Sparta–White County Chamber of Commerce
16 West Bockman Way
Sparta, TN 38583
(615) 836-3552

6. Cedars of Lebanon State Park

Destination: Hidden Springs and Limestone Sinks Nature Trail, Cedars of Lebanon State Park
Route: South on S.R. 10 (U.S. 231S); begins at I-40, Exit 238 (Map 19)
Cities and towns: Lebanon, Bairds Mill
Trip length: 8.3 miles one-way; to return to I-40, retrace route
Nature of the roads: Two-lane, paved highways, mostly straight with some curves and hills; park road is paved and two lanes
Beginning elevation: 575 feet
Ending elevation: 685 feet
Special features: Limestone bedrock, sinkholes, caves, cedar glades, largest stand of eastern red cedar trees in U.S.

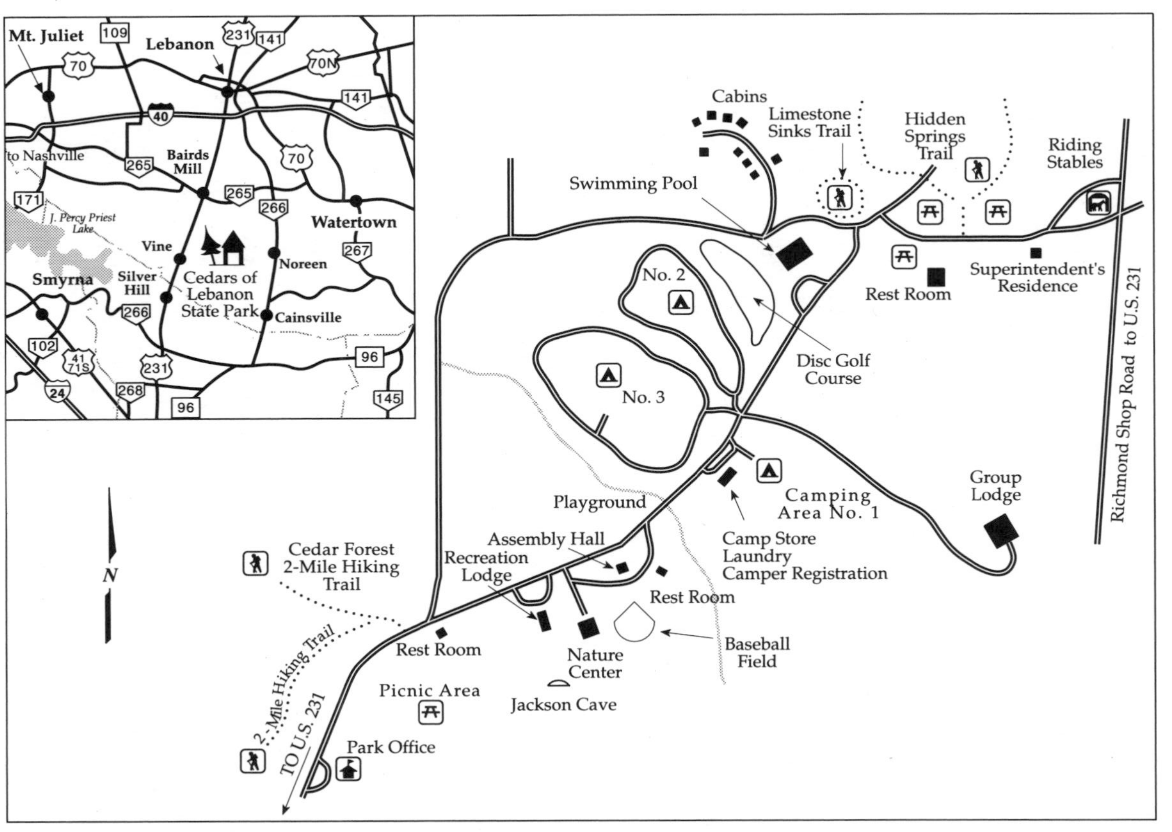

Map 19. Cedars of Lebanon State Day-Use Park. Courtesy of Tennessee Department Conservation and Environment, State Parks Division.

Hiking trail (optional): Hidden Springs Trail or Limestone Sinks Nature Trail
U.S.G.S. 7 1/2 minute quadrangle maps: Gladeville 314 SW, Vine 314 SE
For reference: Central Basin, Ordovician age, Lebanon and Ridley formations, karst, sinkholes, cedar glades

Directions. From I-40 take Exit 238. Turn south (left) onto S.R. 10 (U.S. 231S). Go south on S.R. 10 for 6.6 miles; turn left at Cedars of Lebanon State Park entrance and follow the park road approximately 0.3 mile to the park office and visitors information center.

Route details. Traveling the short distance from the interstate to the entrance of Cedars of Lebanon State Park, you will see the mixture of flatlands and low hills that characterize the Central Basin, the flat-lying limestone bedrock, and numerous eastern red cedar trees.

The route begins along a very straight, flat section of S.R. 10. At Log Mile 2.5 the road climbs a prominent hill, where the flat-lying Ordovician age limestone can easily be seen along the road. At Log Mile 3.6 you can see the operations of a limestone quarry along both sides of the highway. The Ordovician limestone, quarried in open pits, is used for an assortment of crushed limestone products. Most of the crushed limestone gravel goes into driveways, highway construction, and other types of construction.

As you approach the entrance to Cedars of Lebanon State Park, a thin, gray, slabby rock is noticeable in outcrops along the road. This is the Ordovician age Lebanon Formation which plays an important role in the formation of the cedar glades found in this area.

The entrance to Cedars of Lebanon State Park is at Log Mile 6.6. The park got its name from the densely occurring cedar trees, which reminded early settlers of the Biblical land of Lebanon. These cedars, covering many hundreds of the park's 8,887 acres, constitute the largest red cedar forest remaining in the United States. Actually, the eastern red cedar is not a true cedar but a juniper adept at populating areas of very thin, poor soil and rocky conditions.

The rock formation that typically occurs in the cedar glades is a thin-bedded limestone of Ordovician age. Originally called the "glade limestone" by Safford in 1869, this thin, slabby limestone is now referred to as the Lebanon Limestone (C. W. Wilson 1980: 4). Weathering of the Lebanon Limestone typically produces glades, areas of thin to nonexistent soil with thin slabs of the shaly limestone littering the ground surface. The eastern red cedar as well as many wildflowers, mosses, and lichens are well adapted to surviving in these environments. The prickly pear cactus, for example, is a common plant species found in the desert-like glade areas.

Once in the park you notice the dominance of the cedar trees, interrupted only occasionally by deciduous hardwoods and scrubby timber. A stop at the park office and visitors center (approximately 0.3 mile from the entrance) will provide you an opportunity to obtain maps as well as information about hiking trails, horse trails, camping, and lodging. The park's facilities include a swimming pool, picnic areas, a nature center, and a group lodge.

If you lack the time to do the optional half-day hike of the Hidden Springs Trail described below, the Limestone Sinks Self-guiding Nature Trail is an excellent choice for a quick introduction to the environment at Cedars of Lebanon State Park. This informative loop trail, about 0.5 mile long, begins at Log Mile 8.2 (this is along the park road, about 0.1 mile before you get to the Hidden Springs Trailhead). On this trail you can see various geologic features, such as sinkholes and cave openings. The numbered markers along the Limestone Sinks Trail, corresponding to descriptions in a pamphlet provided by the park visitors center, helpfully correlate the living environment—trees, flowers, and animals—with the geology.

An unusual aspect of the park geology is the karst landscape. Areas underlain by limestone and dolostone weather to produce surface features such as depressions, sinkholes, dry creekbeds, and cave entrances (Fig. 121). Such features are produced by the chemical weathering of the limestone or dolostone bedrock, which takes place along joints or cracks in the rock strata where the slightly acid groundwater dissolves the limestone bedrock. As these solution cavities enlarge, they interconnect, providing a network of channels by which the infiltrating surface water can flow.

Over a period of time these subsurface solution cavities interconnect with the surface landscape, producing sinkholes, caves, and an absence of surface streams. A number of caves are found in the park area; Jackson Cave, for example, is located behind the Dixon Merritt Nature Center, approximately 0.6 mile from the visitors center (Fig. 122). Wilson (1980: 13–16) discusses 18 different caves that have been found in the park, giving descriptions of individual caves, and Barr (1961:

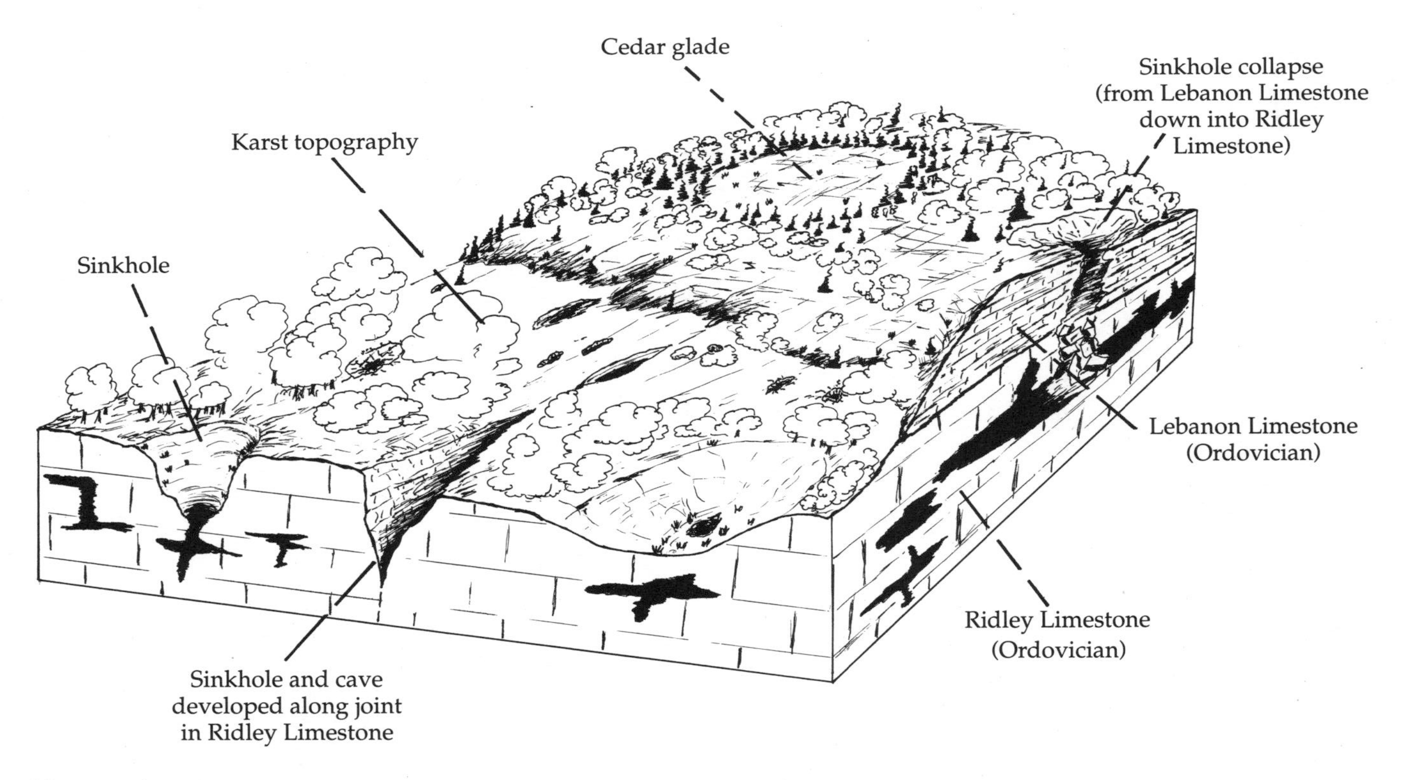

Fig. 121. This schematic drawing illustrates the karst features found at Cedars of Lebanon State Park.

Fig. 122. Jackson Cave, located behind the Merit Nature Center, is one of many caves found in the Ordovician-age limestones of Cedars of Lebanon State Park.

525–27) gives a good description of Jackson Cave. Before entering any park cave, contact the park headquarters for information and permission.

Caution: Because a number of these caves receive surface drainage, in times of heavy rain they can fill up with water, possibly trapping the caver. Remember, never go cave exploring by yourself, always take three sources of light, wear protective clothing and have safety equipment, and tell someone where you are going and when you plan to return.

After leaving the park office, you turn right onto the park road and follow the signs to the Hidden Springs trailhead, approximately 1.4 miles ahead at Log Mile 8.3.

Hidden Springs Hiking Trail (optional)

Destination: Hidden Springs
Location: Trail begins and ends along a paved road in Cedars of Lebanon State Park, approximately 1.4 miles from park office (follow signs)
Start: Hidden Springs Trailhead Parking Area, elevation 575 feet
End: Hidden Springs, elevation 685 feet (then trail loops back to starting point)
Nature of the trail: Unpaved foot trail, moderately easy; flat to rolling terrain; very rocky in places; muddy in wet weather
Distance: 5-mile loop
Special features: Karst landscape (caves, sinkholes, subsurface streams), numerous limestone exposures, flat-lying sedimentary rocks, cedar glades, wildflowers

This hiking trail takes you across a most unusual landscape with numerous depressions, sinkholes, cave openings, rock outcroppings, and cedar trees. The loop trail, 5 miles round-trip, makes a most enjoyable half-day outing. The trail takes you not only by limestone outcroppings and sinkholes, but also across numerous cedar glades. An interesting feature of the glades is how the plants and animals have adapted to this specialized environment.

Notable geologic features of this trail are the limestone rock strata, with their unusual weathering characteristics, and the karst topography. The glades are formed in the thin-bedded, shaly Lebanon Limestone, a formation that itself does not tend to produce karst topography. It is the underlying Ridley Formation, another limestone, that is prone to develop sinkholes and other karst features.

It is believed that the Ridley Limestone forms much of the karst (e.g., cave systems, underground drainage, sinkholes) found in the area (C. W. Wilson 1980: 9). However, in places where the Lebanon Limestone overlies some of the more karstic areas, the subsurface karst features of the Ridley Limestone extend up into the Lebanon Limestone, being expressed as sinkholes in the Lebanon Formation. In other words, as caves and sinkholes develop in the Ridley Limestone, they continue to enlarge toward the surface and eventually collapse some of the overlying Lebanon Limestone.

Particular karst features to look for include dolines (sinkholes), swallets (small cave openings in dolines), and ponors (sinkholes or caves into which surface streams drain), as well as caves and springs, some of which may be hidden beneath the surface.

Trail description. The Hidden Springs Trail, a 5-mile loop, traverses a number of landscapes in the park. Although you may go in either direction, the trail description here takes you to the left and follows the trail clockwise back to the starting point. Both directions are marked with a white paint trail marker (ask at park headquarters for a map with current markings).

Soon after you cross the road from the parking area, you encounter some medium gray limestone outcrops and sinkholes. After passing these outcrops, you cross a gravel road and reenter the cedar forest. The first several hundred yards of the trail border the Limestone Sinks Nature Trail, the self-guiding trail mentioned earlier. (*Note:* The Hidden Springs Trail goes near the other trail but does not connect with it.) Numerous sinkholes, depressions, and cave openings—and the omnipresent cedar trees—are among the features you can observe in this area (Fig. 123).

At about 0.5 mile there is a noticeable change in the forest cover and forest canopy as the trail enters a short section of deciduous hardwood

Fig. 123. The trail to Hidden Springs passes through karst landscape where you will see many sinkholes, depressions, and cave openings.

forest. The trail reenters the cedars after about 100 yards, quickly encountering sinkholes again. At about 0.75 mile the trail crosses several large depressions, each about an acre in size. Notice where there are intersections with horse trails: be sure you stay on the more narrow footpath. After about 1.2 miles you come upon several open glade areas. These glades are characterized by bare patches of a shale-like, slabby limestone (Fig. 124). Wildflowers, some sparse grasses, prickly pear cactuses, and lichens (reindeer-moss) are found in these glade environments.

The trail crosses another dirt road and horse trail and climbs gently to a high point, bordering deciduous forests and nearby pastures. After approximately 2.3 miles you come upon a fenced pit in the bedrock. If you listen carefully, you can detect the sound of running water beneath the trail. This is Hidden Springs. The springs are located about 30 feet beneath the surface in a cave passage (Fig. 125). *Caution:* Be careful not to slip around the open pit, and be sure to control young children.

Notice the dry creekbed floored with medium gray limestone of the Lebanon Formation. This is a good example of karst stream piracy: an

Fig. 124. Weathering of the shaly Lebanon Limestone produces many bare rock exposures where cedar trees flourish. These areas are called cedar glades and are numerous in Cedars of Lebanon State Park.

Fig. 125. Hidden Springs is located down a vertical pit in cavernous limestone. The opening is fenced; caution is strongly urged, especially if young children are present.

underlying cave passage has robbed the surface stream, leaving the streambed dry except when there is rain (Fig. 126). Continuing on the trail about 0.1 mile past Hidden Springs, you encounter a large cave entrance to the left of the trail. This is a ponor, the place where the surface stream once entered the cave system. *Caution:* Be very careful around the edge of the sinkhole—there is a vertical drop of 20 to 25 feet!

The trail soon enters a wide, flat area full of cedar trees and small sinkholes and depressions. For the next 1.5 miles you will encounter many cedar glades, broken only by pockets of small sinkholes. Here you will experience the true nature of the cedar glade environment.

The Lebanon Limestone weathers to a shallow soil of low fertility and produces numerous bare rock outcroppings. A peculiar little plant growing on these glades is reindeer-moss, a shrubby lichen that forms large patches on rock surfaces or thin-soiled areas. This lichen, slow-growing and long-lived, is a sun-loving plant that flourishes in cold, dry climates. Here in the open glades the environment can be very harsh: drought and extremes in temperature are common (Reeves and Somers 1991: 11–14).

Fig. 126. Numerous dry creek beds characterize the karst of Cedars of Lebanon State Park, where few surface streams exist. Most surface water flows into sinkholes where it enters cave systems and becomes part of the groundwater system.

You will notice that along the trail some of the shallow depressions fill up with rainwater during the wet seasons of winter and spring. These water-filled depressions support small ecosystems characterized by algae, invertebrate organisms, turtles, and frogs. After about 4 miles the trail enters an area of mixed deciduous trees and cedars, with an occasional sinkhole and depression. Near the end of the trail you will cross yet another shallow but large depression, which reflects the extensive nature of the karst. At 5 miles the trail joins the trailhead, and the parking area is visible across the road.

Additional information (lodging, food, schedules of events):

Cedars of Lebanon State Park
Superintendent's Office
328 Cedar Forest Road
Lebanon, TN 37087
(615) 443-2769

Lebanon–Wilson County Chamber of Commerce
149 Public Square
Lebanon, TN 37087
(615) 444-5503

7. Wells Creek Structure and Dunbar Cave State Natural Area

Destination: Dunbar Cave State Natural Area, Montgomery County, via Wells Creek structure, Houston and Stewart counties
Route: S.R. 13, S.R. 149; begins at I-40, Exit 143 (Map 20)
Cities and towns: Waverly, Cumberland City, Clarksville
Trip length: 65.8 miles one-way to Dunbar Cave State Natural Area; to return to I-40, retrace the route or take shorter route back to Nashville via I-24 (32.1 miles to Davidson County line, 48.1 miles to downtown Nashville)
Nature of the roads: two-lane, paved highways; mostly straight but some rolling hills; some short curvy sections; four-lane roads in Clarksville
Beginning elevation: ± 455 feet
Ending elevation: ± 405 feet
Special features: Rolling landscape of Western Highland Rim, Wells Creek cryptoexplosive geologic structure, flat-lying sedimentary rocks, cave and karst environment, Mississippian age rocks
Hiking trail (optional): Dunbar Cave
U.S.G.S. 7 1/2 minute quadrangle maps: Clarksville 301 SE
For reference: Western Highland Rim, Mississippian age, caves, karst, cryptoexplosive structure

Directions. From I-40 West take Exit 143. Turn right (north) onto S.R. 13 and go north on it for 34.7 miles. Turn left onto S.R. 49 and go 0.2 mile; then turn right onto S.R. 149 and go 23.5 miles to S.R. 13. Turn left onto S.R. 13 and follow it for 3 miles; then turn left onto U.S. 41A North (S.R. 13 and S.R. 12) and follow it for 3 miles. Turn left onto S.R. 13 and go 1.4 miles to the sign for Dunbar Cave State Natural Area. Turn right onto Dunbar Cave Road and go 1 mile; the park office is on the left.

Route details. Traveling along this route, you will see the rolling countryside of the Western Highland Rim, pass through the Wells Creek structure near the Cumberland Steam Plant, cross the Duck and Cumberland rivers, and view the karst landscape around Clarksville, Tennessee.

The route begins along the western slopes of the Western Highland Rim, where erosion has etched the landscape with numerous hollows and the valley of the Duck River (Log Mile 4.2). Exposed bedrock is not common along the route. Along the roadway cutslopes and in adjacent woods and pastures, you will see the reddish-orange residual soil of

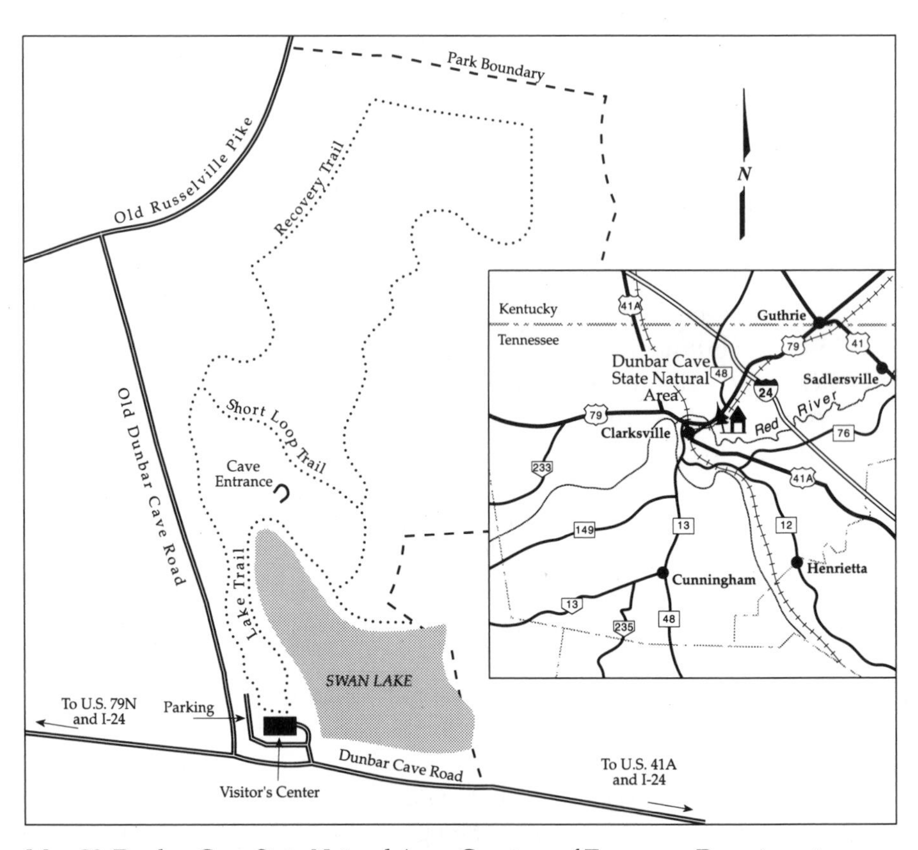

Map 20. Dunbar Cave State Natural Area. Courtesy of Tennessee Department Conservation and Environment, State Parks Division.

the Mississippian limestone. Mixed in the soil are fragments of chert, which weather to a color ranging from white to buff-brown and have a gravel-like appearance.

The route goes through Waverly, passing the Humphreys County Courthouse and continuing north toward Erin. In Houston County at Log Mile 29.7, the route crosses Tennessee Ridge, the divide between the drainage area of the Tennessee River and that of the Cumberland River. This area is underlain by relatively flat-lying Mississippian limestone. At Erin the route leaves S.R. 13 and turns onto S.R. 49 for just 0.2 mile (Log Mile 34.7–34.9), then turns right (north) onto S.R. 149. At a point about 0.1 mile onto S.R. 149 (Log Mile 35.0), you can see the first influence and exposures of the Wells Creek structure: a large bluff of Mississippian age limestone to the left of the highway. If you look carefully, you can see a slight dip in the layers of rock. This dip, away from the central point of the Wells Creek structure, indicates that tremendous geologic forces (possibly a meteor) have deformed the originally flat-lying limestone.

Continuing north, the route crosses over weathered strata of the Wells Creek structure, ranging in age from Mississippian down to Cambrian. Near Log Mile 39 the route crosses near the central section of the structure, represented by several rolling, grassy hills to the left (west) of the highway, just inside the Stewart County line. The bedrock of the area and the Wells Creek structure itself are not easily discerned from the highway. Indeed, only on close examination could a geologist discover the true nature of the Wells Creek structure.

The Wells Creek structure (Fig. 127), with a diameter of approximately 8 miles, is a cryptoexplosive feature (Wilson and Stearns 1968: 1–17, 163–77; Miller 1974: 55–58; Price 1991: 22–26). Generally circular in structure, it was formed in some manner from a natural, explosive release of energy (Price 1991: 24). Most geologists believe that this structure resulted from the impact of a meteor—a crashing, cataclysmic impact sending the intruding material at least 2,000 feet beneath the surface of the ground. The violent explosion shattered the rock layers, producing "shatter cone" structures within the central core of the rock mass, which is limestone of the Knox Formation.

Shatter cones are conical fragments of rock, usually 1 to 5 inches long and about the size of a fist, that are distinctively striated with fracture lines. The tip of the shatter cone points in the direction of the applied pressure. In the case of the Wells Creek structure, it is thought that a

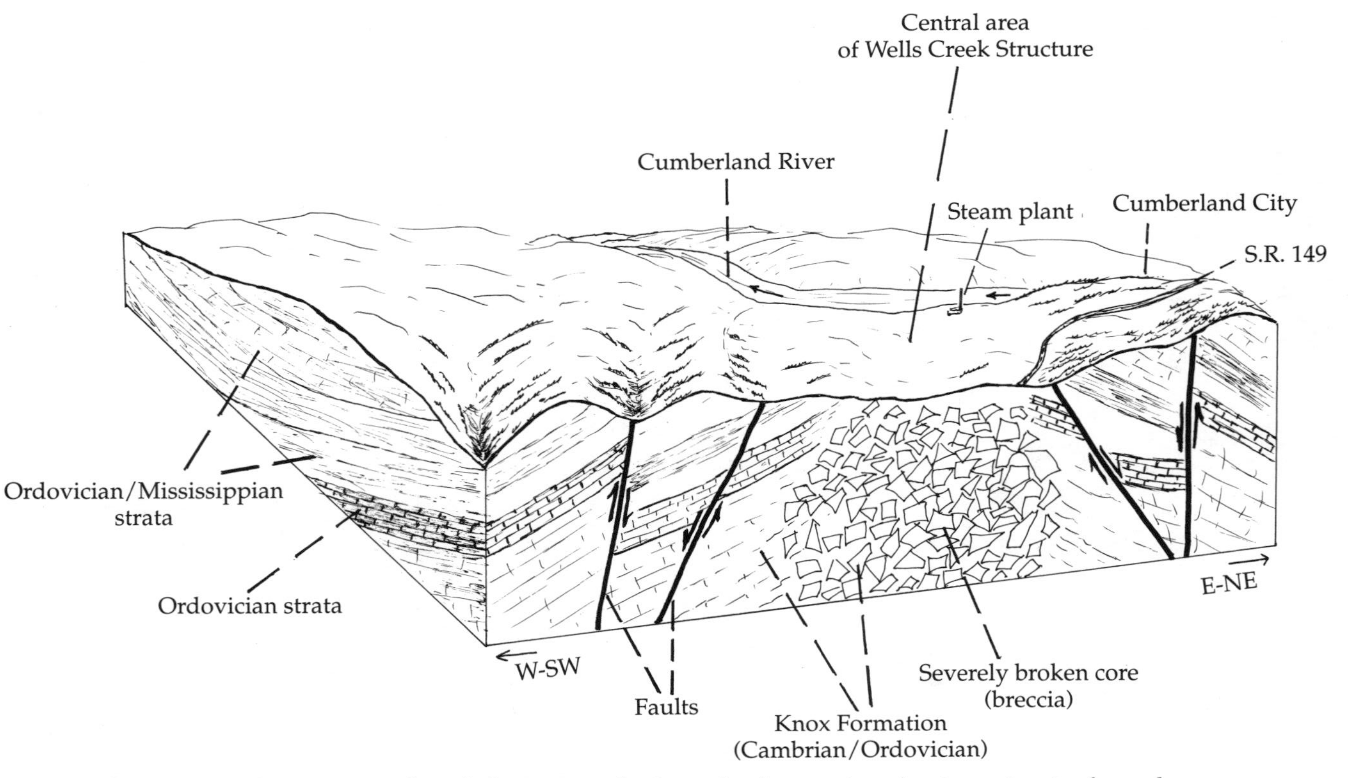

Fig. 127. The Wells Creek structure is thought by most geologists to be the remains of an impact crater formed many thousands of years ago by the concussion of a meteorite or possibly a comet.

central point of explosion was about 2,000 feet underground at the time of the event (Wilson and Stearns 1968: 130). The resulting explosive energy tilted the sedimentary strata upwards from the central impact area and broke the rock layers along circular fault paths.

Subsequent weathering and erosion has removed possibly several hundred feet of rock material, mostly in the shattered central impact area (approximately 2 miles in diameter). Today's landscape reveals only rolling, hilly terrain, few good examples of natural outcrops of bedrock, and (if you use your imagination) a somewhat basin-like structure.

Another theory about the origin of the structure involves the explosive release of superheated groundwater as a result of deep-seated volcanic activity. Perhaps upwellings of molten rock in the crust approached near the surface crust but did not extend through the local sedimentary rock. Heat from the molten material raised the temperature of nearby areas of groundwater, producing steam. The explosive action of the gas-jet supercritical steam (water vapor under high pressure) that broke through the surface shattered and tilted the rock strata from beneath the surface (Wilson and Stearns 1968: 166–68).

A possible explanation for the deep-seated energy source might be "hot spots," places in the earth's crust where thinning has permitted mantle material to rise near the outer crust, creating a spot of geothermal activity and possibly volcanoes. As the crust was being pulled across a hot spot, the cryptoexplosive activity could have occurred.

The lack of igneous rocks, even beneath the near surface, and the lack of heavy mineralization from the enriched supercritical steam tend to discredit the cryptoexplosive theory. However, the lack of meteorite material likewise undermines the meteorite theory. All in all, the origin of the Wells Creek structure is still a mystery, although most geologists tend to be swayed by the meteorite impact theory.

At Log Mile 39.4 there is a turnoff (left) to Cumberland Steam Plant (Fig. 128) and downtown Cumberland City, but the route continues north on S.R. 149, passing east of Cumberland City. As you turn northeast, you cross a rolling to hilly section of the Highland Rim. At Log Mile 43.4 the highway crosses from Stewart County into Montgomery County. Along this section of the route, you will see very few rock exposures, with the exception of the road cut at Log Mile 51.1, where Mississippian limestone is visible.

The route, following S.R. 149, intersects S.R. 13 at Log Mile 57.4. Here you turn left (north) onto S.R. 13 and proceed toward Clarksville. At

Fig. 128. Smokestacks from the Cumberland Steam Plant mark the landscape near the center of the Wells Creek structure, where strata of the Cambrian-Ordovician-age Knox Group are found.

Fig. 129. Near Log Mile 57.9 the route of Side Trip 7 crosses the Cumberland River (here flowing from right to left) near Clarksville, Tennessee.

Log Mile 57.9, S.R. 13 crosses the Cumberland River, which you see flowing from right to left (east to west) as you travel north into Clarksville (Fig. 129).

At this point the route enters the urban area, with many side streets and intersections. At Log Mile 60.4, turn left onto U.S. 41A North and follow it for approximately 3 miles. Then, at Log Mile 63.4, you turn left and take S.R. 13 north toward Dunbar Cave State Natural Area.

After crossing the Red River, the route continues up a long hill. At Log Mile 64.8 (about 1 mile beyond Red River), you turn right onto Dunbar Cave Road. Look for the Dunbar Cave State Natural Area sign. Go approximately 1 mile, and you will find the park entrance on the left (Log Mile 65.8).

Dunbar Cave Short Loop Hiking Trail (optional)

Destination: Entrance area of Dunbar Cave
Location: Trail begins at the park office and visitors center at Dunbar Cave State Natural Area, Dunbar Cave Road, Clarksville, Tennessee
Start: Trailhead is at the park office, elevation ± 405 feet
End: Dunbar Cave entrance, elevation 425 feet
Nature of the trail: Partly paved and level to unpaved and hilly; hilly section is moderately difficult
Distance: Paved trail only, 0.66 mile round-trip; paved trail and short loop foot trail combined, 0.8 mile round-trip
Special features: Scenic setting of Dunbar Cave entrance, flat-lying Mississippian age limestone strata, karst landscape

Dunbar Cave State Natural Area is in a karst region of the Highland Rim in Montgomery County. The Dunbar Cave site, covering 110 acres, was designated a State Natural Area in 1973 by the Tennessee General Assembly.

The history of Dunbar Cave is long and varied, extending back in time to the period of Tennessee's prehistoric Indians. Indeed, a Tennessee State Parks brochure on the cave points out that recent excavations in the cave entrance show it to be one of the most potentially significant archaeological sites in the southeastern United States. In the 1770s, Anglo-Europeans discovered the cave, and thereafter the huge cave entrance has been used as a center for community social gatherings. At the beginning of the twentieth century, the Dunbar Cave site was a resort with mineral springs. In the

1930s the cave was popular for dances with many of the famous Big Bands. In 1948 country music star Roy Acuff purchased the cave site, using it for weekend square dances and country music shows. He sold the property in 1963, and the state purchased it in 1973 (Sellers 1991).

Dunbar Cave is presently known to contain about 10 miles of explored and mapped passageways. It is inhabited by the blind cave fish *Typhlichthys subterraneous,* which are unpigmented; they are the subject of numerous cave biology studies (Barr 1961: 331).

Cave hikes are conducted by park personnel for organized groups (10 to 20 persons) by appointment. On Saturdays and Sundays during June, July, and August, hikes are also conducted for the general public. The hike to the entrance and the short loop trail are very leisurely walks where you can enjoy not only the geology but wildflowers and a scenic lake.

Trail description. The trail begins at the visitors center and takes you along the edge of the park's lake (Swan Lake). This section of the trail, called the Lake Trail, is a paved, relatively level walkway that leads you in 0.33 mile directly to the entrance of Dunbar Cave (Fig. 130).

Fig. 130. The trail to the entrance of Dunbar Cave passes along the shore of the artificial Swan Lake.

At the cave entrance you can observe the 50-foot-high bluff sheltering the natural entrance, which measures 35 feet wide and 10 feet high (Fig. 131). The locked gate at the entrance is necessary not only to protect the cave and its life forms from unnecessary human intrusions but also to prevent unauthorized expeditions that might be dangerous for visitors without a guide.

Dunbar Cave has formed in flat-lying, Mississippian age limestone of the St. Louis Formation (Fig. 132). Cave passages have developed mainly along joints which trend north-northwest. The main trunk galleries, or passageways, average 15 to 20 feet in width and 8 to 12 feet in height. Several small pits connect different levels of the cave, providing access to upper, older levels and younger, lower, wet levels (Barr 1961: 331).

The development of this cave is the result of the chemical weathering of the limestone bedrock: this limestone goes into solution when it comes in contact with the groundwater, which is naturally slightly acidic. The landscape of the surrounding area is characterized by karst features—sinkholes, depressions, "sinking" streams, and caves. The process of cave development is still active in the Clarksville area, as the

Fig. 131. Dunbar Cave is developed in flat-lying Mississippian-age limestone as seen here at the cave entrance.

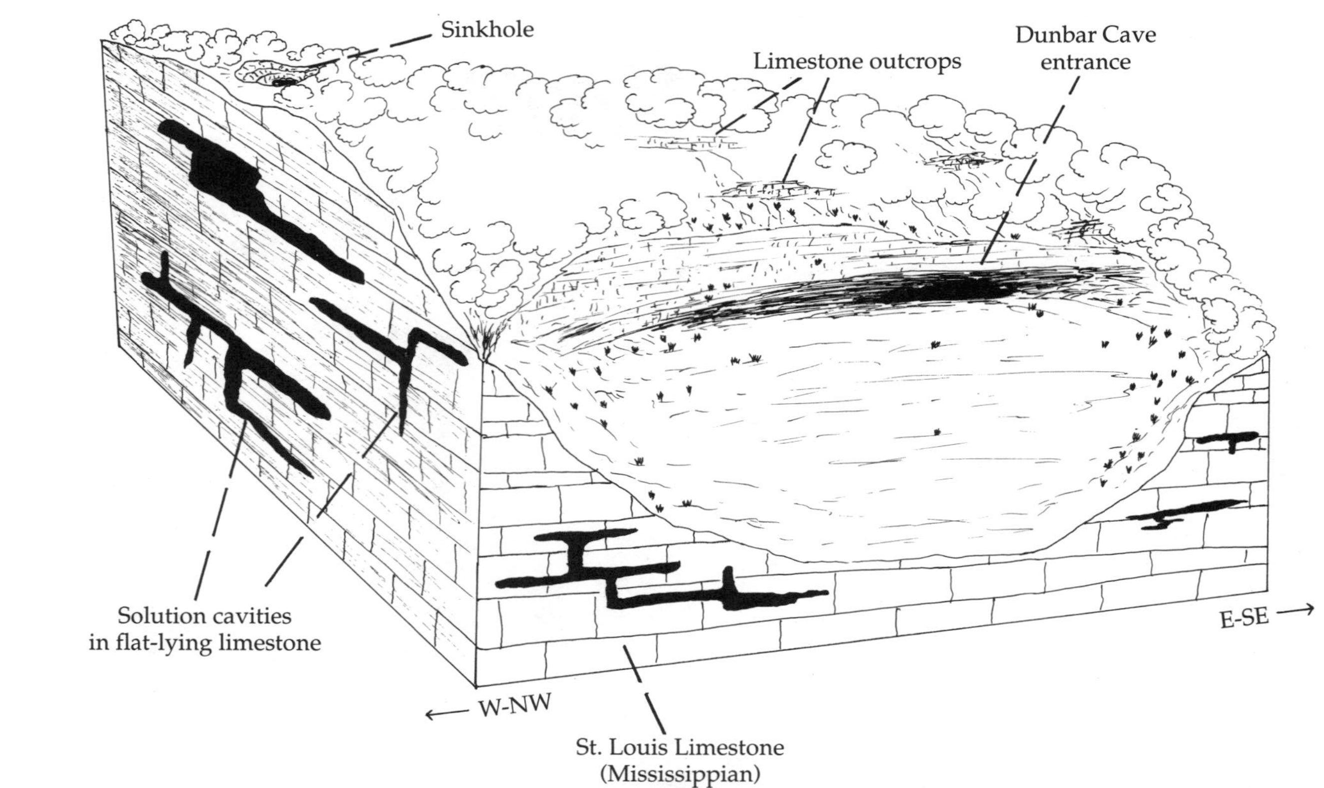

Fig. 132. Dunbar Cave is formed in flat-lying Mississippian-age limestones that have undergone extensive chemical-solution weathering.

numerous sinkholes and depressions there attest. Active sinkhole collapse, a major problem in this urban area, requires much study and analysis (Kemmerly 1980: 3–5).

From the cave entrance area, the trail goes beneath the entrance patio and continues along the lakeshore. After about 100 yards the paved trail ends, and the unpaved short loop trail begins, turning left up the hill. The trail becomes moderately steep and rocky as it climbs the ridge above the cave entrance.

After about 250 yards the loop trail turns left again, near the top of the hill, and begins a gentle descent off the west side of the hill. You can see several sinkholes and depressions that serve to recharge the cave system with fresh rainwater. In the spring there are numerous wildflowers along this section of trail, including trillium and may-apple.

Now the trail quickly descends the hill, looping back to the cave entrance. From this point you return to the visitors center along the Lake Trail.

Additional information (food, lodging, schedules of events):

Dunbar Cave State Natural Area
401 Dunbar Cave Road
Clarksville, TN 37043
(615) 648-5526

Clarksville Area Chamber of Commerce
312 Madison Street
P.O. Box 883
Clarksville, TN 37041-0883
(615) 647-2331

Humphreys County Chamber of Commerce
124 East Main
P.O. Box 733
Waverly, TN 37185
(615) 296-4865

Stewart County Chamber of Commerce
363 Spring Street, Watson Building
P.O. Box 147
Dover, TN 37058
(615) 232-8290

8. Reelfoot Lake State Park

Destination: Reelfoot Lake State Park
Route: U.S. 412, S.R. 78 and S.R. 22 (for additional trip around Reelfoot Lake, S.R. 22 and S.R. 157; Kentucky Route 311, an unnamed county road, Ky. Rt. 94, and S.R. 78); begins at I-40, Exit 79 (Map 21)
Cities and towns: Jackson, Alamo, Dyersburg, Bogota, Tiptonville
Trip length: 69.6 miles one-way to Reelfoot Lake Visitors Center; optional drive around Reelfoot Lake is additional 40 miles (circular route beginning at visitors center)
Nature of the roads: Four-lane and two-lane paved highways; mostly rolling to flat, with few curvy sections
Beginning elevation: ± 350 feet
Ending elevation: ± 285 feet
Special features: Gulf Coastal Plain, Mississippi River alluvial floodplain, loess deposits, Jackson Formation, Reelfoot Lake, Reelfoot Lake National Wildlife Refuge
Hiking trail (optional): Boardwalk at Visitors Center
U.S.G.S. 7 1/2 minute quadrangle maps: Tiptonville 419 NW, Ridgely 419 SW, Hornbeak 419 SE, Samburg 419 NE
For reference: Gulf Coastal Plain, Reelfoot Lake, New Madrid fault

Directions. From I-40 take Exit 79 onto U.S. 412. Turn right onto U.S. 412 and follow it through Madison, Crockett, and Dyer counties until the intersection with I-155 in Dyersburg, where you take the ramp onto I-155 West. Follow I-155 to Exit 13; there you exit I-155 and turn right onto S.R. 78, which you follow through portions of Dyer, Obion, and Lake counties into Tiptonville. Turn right from S.R. 78 onto S.R. 22 and follow it 1.4 miles to Reelfoot Lake State Park Visitors Center.

Geologic and Environmental Significance

The side trip to Reelfoot Lake provides a look at both the topography of the Gulf Coastal Plain and Mississippi River Alluvial Floodplain and the nature of the impact of the earthquakes on the landscape of Northwest Tennessee. Intensive agriculture across West Tennessee has led to several environmental concerns, one of which is directly related to the geology. The practice of draining swamplands (often referred to as wetlands) for agricultural use has developed into a major point of concern between the farmers, who want to use the land to raise crops,

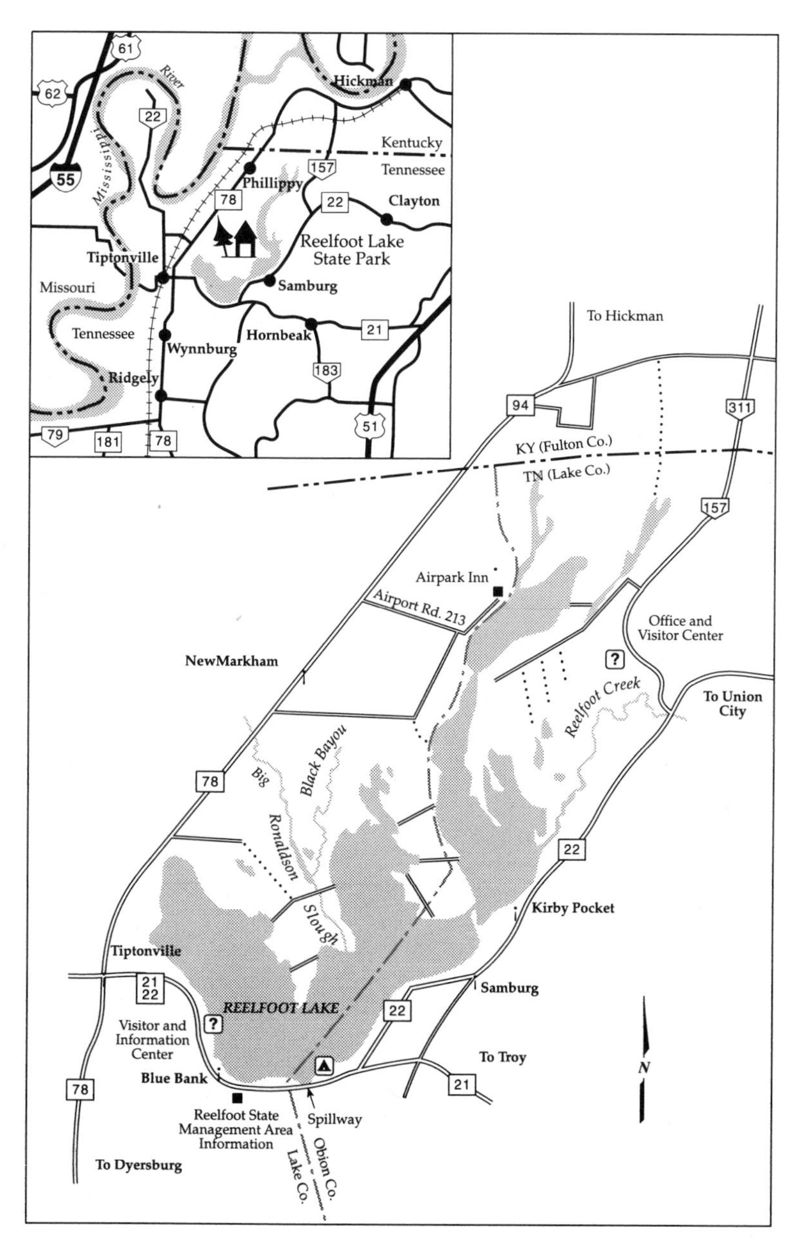

Map 21. Reelfoot Lake State Park. Courtesy of Tennessee Department Conservation and Environment, State Parks Division.

and the environmental groups, who want to protect the habitat of wild plants and animals. Some of the wetlands were farmed in earlier days but were thereafter abandoned and left to redevelop as wetlands.

This issue is further complicated by the problem of siltation. As farmers plow, disk, and break up the ground in order to plant and raise a crop, the soil becomes loosely compacted. Most of the soil in West Tennessee is composed of loess, an ancient wind-deposited silt that erodes very slowly when exposed on nearly vertical slopes, such as bluffs, but quickly erodes on flatter slopes. As the ground is broken up in order to plant crops, rain rapidly erodes the bare ground, carrying the silt to nearby ditches, gullies, and rivers. This siltation quickly builds up in the streams, choking them and altering their courses.

Recent developments in agricultural practices have helped reduce the amount of soil eroded from the land in West Tennessee. Perhaps the most unusual of these techniques is no-till planting. Instead of plowing and tilling the fields, the farmer simply plants the seed, using a machine that disturbs the soil very little. The result is far less loose soil and much less potential for erosion. Another technique is the construction of small, shallow siltation ponds along the lower slopes of plowed fields. Although the ponds do not prevent erosion, they collect the silt-laden runoff from fields and prevent it from polluting streams. Along with these newer techniques, some older ones are being reemphasized, including contour plowing and crop rotation. With contour planting, furrows flow around a hill at constant height rather than up and down, so water is not encouraged to rush down the hill, carrying silt with it. Crop rotation tends to alternate row crops and pasture use, so that the land has a chance to stabilize itself rather than be constantly exposed to disturbance.

As you travel across West Tennessee from Jackson to Reelfoot Lake, you will find that the silt loess deposits increase in thickness from a feather's edge at Jackson to over 50 feet along the bluffs near the Mississippi River floodplain. These loess bluffs mark the boundary between the Gulf Coastal Plain province and the Mississippi River Alluvial Floodplain province. This boundary is located at Log Mile 50.7 on S.R. 78 in Dyer County, just as the highway crosses the Obion River.

At Reelfoot Lake, located on the Mississippi River floodplain near the base of the loess bluffs, the siltation problem is rapidly contributing to the filling in of the natural lake (Denton 1986: 30–35). This silt has its origins both along the loess bluffs and in the plowed fields beyond the bluffs. Local and regional soil conservation programs have reduced the rate of the silting in of Reelfoot Lake, but the problem still exists. One plan considered by local groups entailed the temporary draining of the lake in order to dredge out the silt deposits (McIntire, Naney, and Lance 1986: 14–20). Certain aspects of this option, however, made it undesirable: the temporary loss of aquatic habitat; detrimental effects on the populations of not only waterfowl such as herons and ducks but also hawks and, of course, the bald eagles; and a consequent loss of tourism for the lake area. Accordingly, the lake was not drained and dredged.

The Geologic Origin of Reelfoot Lake

The area in and around the Reelfoot Lake vicinity, and along the Mississippi River from Ohio and Illinois down to the Gulf of Mexico, is known as the Mississippi Embayment. During the Mesozoic and Cenozoic eras, it was covered by the Gulf of Mexico. Sediments accumulated in the area during the end of the Mesozoic era (Cretaceous period), continuing into the Cenozoic era. The thickest accumulation of sediments occurred along the long axis of the embayment (a trough-shaped basin), which is along the present position of the Mississippi River. One of the locations of these thick sediments—the New Madrid, Missouri, area—corresponds to an area of active seismicity.

Although these thick sediments persist southward toward the present Gulf of Mexico, the seismic activity is situated along the New Madrid fault zone. This area, parallel to the Mississippi River in Tennessee, Arkansas, and Missouri, extends from near Memphis to just north of New Madrid, Missouri. Deep within the crust beneath the area lies a major fault, the New Madrid, with numerous smaller associated faults.

Subsidence of the embayment area was the result of intracontinental rifting (faulting) that began in late Precambrian or early and middle Cambrian time (Obermeier 1989: 7). During the Paleozoic era, marine sediments were deposited in the newly formed trough. From late in the Mesozoic era (the Cretaceous period) on down into the beginning of the Holocene epoch of the Cenozoic era, subsidence of the embayment reoccurred, and more and more sediments accumulated in the embayment trough.

Recent seismic activity, including the 1811–12 New Madrid earthquakes, is occurring along the rift zone, deep beneath the present-day Mississippi River, in Tennessee, Arkansas, and Missouri.

The measurement of earthquakes. The familiar Richter Scale, numbered from 1 to 10, measures the magnitude of energy released during an earthquake. This scale is based on a logarithm, to the base 10, of the amplitude recorded on a seismogram. On the Richter Scale the 1-unit difference in a magnitude of 4 and 5 represents a wave amplitude on the seismogram 10 times greater but an energy release 32 times greater. The largest earthquakes ever recorded were 8.9 events: in 1906 at the border of Columbia and Ecuador and in 1933 at Sanriku, Japan. The Modified Mercalli Scale utilizes a different system of determining the intensity of an earthquake, one based upon human observations of the damage to natural and artificial objects (Templeton and Spencer 1980: 3–4). There is no accurate way of comparing the two scales, both of which are widely used by scientists as well as the news media: variations in rock type, structure of the beds of rock, and surface conditions, for example, can make two earthquakes of the same magnitude (Richter Scale) have very different intensities (Mercalli Scale).

Between December 1811 and February 1812, over 2,000 earthquakes were generated in the seismically active area along the New Madrid fault zone. The epicenter of the three largest ones was New Madrid, Missouri. On December 16, 1811, an earthquake with a magnitude of XI on the Modified Mercalli Intensity Scale (Wood and Neumann 1931: 277–83) shook not only West Tennessee but Arkansas, Missouri, Kentucky, and Illinois as well (Templeton and Spencer 1980: 13). A little over a month later, on January 23, 1812, a second major quake occurred with an intensity of X, vibrating the farmlands in West Tennessee, Eastern Arkansas, and Southeastern Missouri. On February 7, 1812, another violent earthquake again shook the Reelfoot area with an intensity of XI. All three earthquakes, which had intensities of IX or greater, were felt over an area of 20,000 square miles (Mitchell 1991: 120). It is said that some of the vibrations were strong enough to ring bells in Norfolk, Virginia, and to stop clocks in Boston, Massachusetts.

Movements along the New Madrid fault sent seismic waves throughout the region, shaking the rocks, sediments, and land surface. The sediments near the surface—some of them several thousand feet in thickness—were intensely vibrated (Stearns 1957: 1081–83). In these layers of sand, silt, and clay, the vibrations caused the sand and silt particles to be rearranged, thus allowing the groundwater to liquefy the sandy silt areas. As a result, the overlying sediments collapsed into the liquefied silt and sand, which was then squashed and forced upwards along cracks in the soil. Large fissures opened up, releasing sulfur gas and spewing great plumes of sand and water high into the air. These sand blows, some reaching 60 or 80 feet into the air, dotted much of Eastern Arkansas.

Because of the liquefaction, much of the land surface sank (slumped), causing numerous surface depressions. These depressions, which quickly filled with water from disoriented surface streams, formed ponds and several natural lakes (Reelfoot Lake in Tennessee and St. Francis Lake in Arkansas are the largest and best known). The subsidence of the area now occupied by Reelfoot Lake is evidenced by the presence of many submerged trees.

Slumps also occurred along the loess bluffs, where several of the slump blocks can be seen in the gullies that dissect the bluffs. (A slump is a small-scale landslide in which the soil and rock, in a mass, move in a downward direction, usually with a backward rotation, leaving a semicircular scar on the slope from which they came.) A number of these slump-block landslides probably occurred during the violent shaking of the 1811 and 1812 earthquakes. Others, however, may have occurred during earlier seismic events.

Vivid eyewitness descriptions of the earthquakes and their aftermath are preserved in personal accounts of people living in the region at the time. Juanita Clifton describes some of the scenes: "For those who could see in the night, the earth's surface was like a storm blown ocean, rolling and pitching, rising and falling. Huge cracks appeared in the surface. Some of these were filled and overflowing with sand spews.

"The rolling river was like a maddened beast. The banks could no longer hold it and thousands of tons of water poured out over the land. In its rush to recede into itself, it tore hundreds of trees out by the roots, carrying them into the river. Great depressions covering acres of land appeared and were filled with the black Mississippi waters" (Clifton 1980: 35).

The dramatic events surrounding the genesis of Reelfoot Lake have also been recounted in Chickasaw myth. The only son of a Chickasaw chief was born with a deformed foot. Because the young man walked and ran with a rolling motion, his people called him Kalopin, meaning Reelfoot. When his father died, Reelfoot became chief. In his search for an Indian princess to rule with him, he fell in love with a maiden by the name of Starlight, princess of the Choctaws. After her father forbade her to marry the chief with the deformed foot, Reelfoot and his braves stole Starlight from the Choctaws. The Great Spirit, disapproving of the theft of Starlight, punished Reelfoot and his people. The Great Spirit caused the earth to rock and the waters to swallow up the Chickasaw village, burying the people in a watery tomb. Chief Reelfoot and Starlight, too, were engulfed by the waters. Where the Great Spirit stamped the earth, causing it to shake, the Mississippi River formed a beautiful lake. In the bottom lay Reelfoot, his bride, and his people (P. E. Walker 1929: 15–20).

Although the legend of Reelfoot seems to indicate that Indians inhabited the region, in actuality, very few Indians lived in this part of West Tennessee during this period. Instead, this area was reserved as a hunting ground, which they passed through from time to time (P. E. Walker 1929: 21).

Route details. This route takes you across the rolling to flat landscape of the Gulf Coastal Plain, the flat bottomlands of the Mississippi River Alluvial Floodplain, and to Reelfoot Lake, which was formed by a major earthquake.

Most of the trip takes you into the productive agricultural regions of West Tennessee, with fields of cotton, soybeans, and various vegetables.

Once you leave the interstate, you turn right onto U.S. 412 (Log Mile 0.0) and follow it out of Jackson, quickly finding yourself in rural West Tennessee. At Log Mile 2.5 you enter an area bordering a large, flat floodplain (on your left) of the South Fork of the Forked Deer River. The soil of the floodplain is very fertile, supporting cultivation of a variety of row crops, including cotton and soybeans. At Log Mile 7.5 the route crosses into Crockett County, one of the largest producers of cotton in Tennessee. Continuing in a northwesterly direction across Crockett County, the route intersects U.S. 70 at Log Mile 11.8 (near Bells, Tennessee).

As you travel northwest through Crockett County, the loess deposits gradually thicken. At Log Mile 27.6 the route crosses a prominent ridge near the Crockett-Dyer county line, and at Log Mile 28.1 you enter Dyer County. Numerous bald cypress trees are easily within view to the right of the road as you cross the Forked Deer River at Log Mile 35.7.

U.S. 412 intersects I-155 at Log Mile 40.6. Stay to the left and cross over I-155, following the ramp to I-155 West. Enter I-155 West and stay on it until you reach Exit 13 (approximately 2.3 miles). Leave I-155 at Exit 13 and follow the ramp to S.R. 78. Turn right (north) onto S.R. 78. Reelfoot Lake State Park is 33 miles from this point (Log Mile 43.2).

You will notice that the area is still characterized by rolling hills, which are underlain by the Jackson Formation and capped at the surface with loess deposits. At Log Mile 48.2 the route intersects S.R.

182. From Log Mile 48.2 to Log Mile 50.7, the area becomes very hilly because of the erosion of the loess deposits near the loess bluff escarpment overlooking the Mississippi River Alluvial Floodplain.

At Log Mile 50.2 the route begins the descent from the Gulf Coastal Plain province to the Mississippi River Alluvial Floodplain (Fig. 133). The bluffs on your right are capped with 50 to 80 feet of the windblown loess deposits and underlain by clay, sand, and lignite of the Jackson Formation.

You cross the Obion River and enter the Mississippi River Alluvial Floodplain at Log Mile 51.2. Ahead and to your left you can see the broad, flat plains of the adjacent Obion River floodplain as it overlaps and interfingers with the Quaternary period alluvial deposits of the Mississippi River floodplain.

The route intersects S.R. 103 at the Bogota community (Log Mile 52.8). At Log Mile 54.2 you can see a distant ridge off to the right of the road. This is the edge of the Gulf Coastal Plain province, which is underlain by the Jackson Formation and loess deposits.

At Log Mile 56.8 the route crosses a narrow stretch (about 0.9 mile) of Obion County before entering Lake County at Log Mile 57.7. The landscape now is flat, and the very fertile soil is intensively cultivated.

Fig. 133. The route of Side Trip 8 passes over the boundary between the Coastal Plain and Mississippi River Alluvial Flood Plain near Log Mile 50.7. The view is to the north, where S.R. 78 crosses the Obion River.

The route intersects S.R. 79 at Log Mile 58.1 (you continue on S.R. 78). Shortly you will pass through the communities of Ridgely (Log Mile 60.4) and Wynnburg (Log Mile 64.7). Notice that the highway is built up above the flat floodplain and its fertile alluvial deposits.

You enter Tiptonville at Log Mile 67.7 and quickly reach an intersection with S.R. 21 and S.R. 22 (Log Mile 68.2). Here you should turn right onto S.R. 22. Within a mile after turning onto S.R. 22, you can see the waters of Reelfoot Lake on your left (Fig. 134).

On your left at Log Mile 69.6 is the Reelfoot Lake State Park Visitors Center, which has a museum dealing with the origins of Reelfoot Lake, its wildlife, and the human history of the area. An enclosure next to the museum is the home of bald eagles that were injured and cannot be returned to the wild. The optional hiking trail begins at the rear of the visitors center.

Fig. 134. Reelfoot Lake is characterized by the numerous cypress trees that grow in the shallow lake waters.

Reelfoot Lake Boardwalk Trail (optional)

Location: Trail begins behind Visitors Center
Start and end (loop trail): Reelfoot Lake State Park Visitors Center, elevation ±285 feet; boardwalk on trail goes out onto Reelfoot Lake, lake elevation ±283 feet
Nature of the trail: Flat, paved on land; wooden boardwalk over lake edge
Distance: Approximately 0.25 mile, round-trip
Special features: Cypress trees; shallow lake; wildlife—birds (including bald eagles during winter), turtles, fish; hands-on nature exhibits at visitors center

This short walk introduces you to the Reelfoot Lake environment. You walk among large bald cypress trees, their knees protruding out of the lake waters, and see many varieties of wildlife. Bald eagles, which winter here, can generally be seen around the lake from December through mid-March. Although rocks are not visible here, this is a geologically interesting area. As you look across the water-filled basin, use your imagination to visualize the events of the violent earthquakes of 1811 and 1812 that caused the lake to form.

Trail description. The trail begins at the rear of the visitors center/ museum in a grassy area where you can observe some live bald eagles in the eagle rehabilitation exhibit. Bald eagles, with wingspans of six to eight feet and weights of up to sixteen pounds, annually visit Reelfoot Lake during the winter months. The Reelfoot eagles fly south mainly from the Great Lakes and isolated regions of Canada to winter in a less harsh climate. It is believed that the eagles have visited Reelfoot Lake ever since the New Madrid Earthquake of 1811–12 that created the lake.

The trail continues to the right of the eagle rehabilitation area and enters the boardwalk (Fig. 135). You can go either to the right or to the left: the boardwalk makes a circular route through the trees and out over the water. About midway around the boardwalk there is an observation area.

In addition to the eagles, a wide variety of other birds can be observed at Reelfoot Lake. Over the course of a year, more than 250 species can be found. Some of them occur in large numbers—an estimated 100,000 ducks of various species and 66,000 Canada geese, for example.

Fig. 135. A short trail behind the Reelfoot Lake Visitors Center leads out over the waters of the lake via a boardwalk. Numerous forms of wildlife can be seen from here.

The vast stretch of water before you covers what was once land. As a result of the 1811–12 earthquake, this area experienced the effects of liquefaction, a phenomenon that occurs because of the vibration of saturated silt and sand. When shaken by earthquakes, the water found in this silt and sand beneath the land surface migrates out of the open spaces (interstices) between the individual grains of silt and sand, causing the sand to turn mushy or liquefy. (If you've visited the beach, you may have experienced this phenomenon and not realized it. When you tap your foot on it, the wet sand turns to a mushy liquid.)

As the silt and sand layers beneath the surface were violently shaken, layers of mushy sand formed. When the overlying sediments collapsed upon the weaker liquefied layers of sand and silt, the land surface sank 30 to 40 feet or more. Because the liquefied sand was under pressure from the overlying sediments, it was forced to the surface along cracks in the earth, and plumes of sand (sand blows) were spewed out into the atmosphere. This huge depression, which quickly filled with surface water, became known as Reelfoot Lake (Fig. 136). To return to the Visitors Center, finish the loop around the boardwalk back to the beginning of the trail at the visitors center.

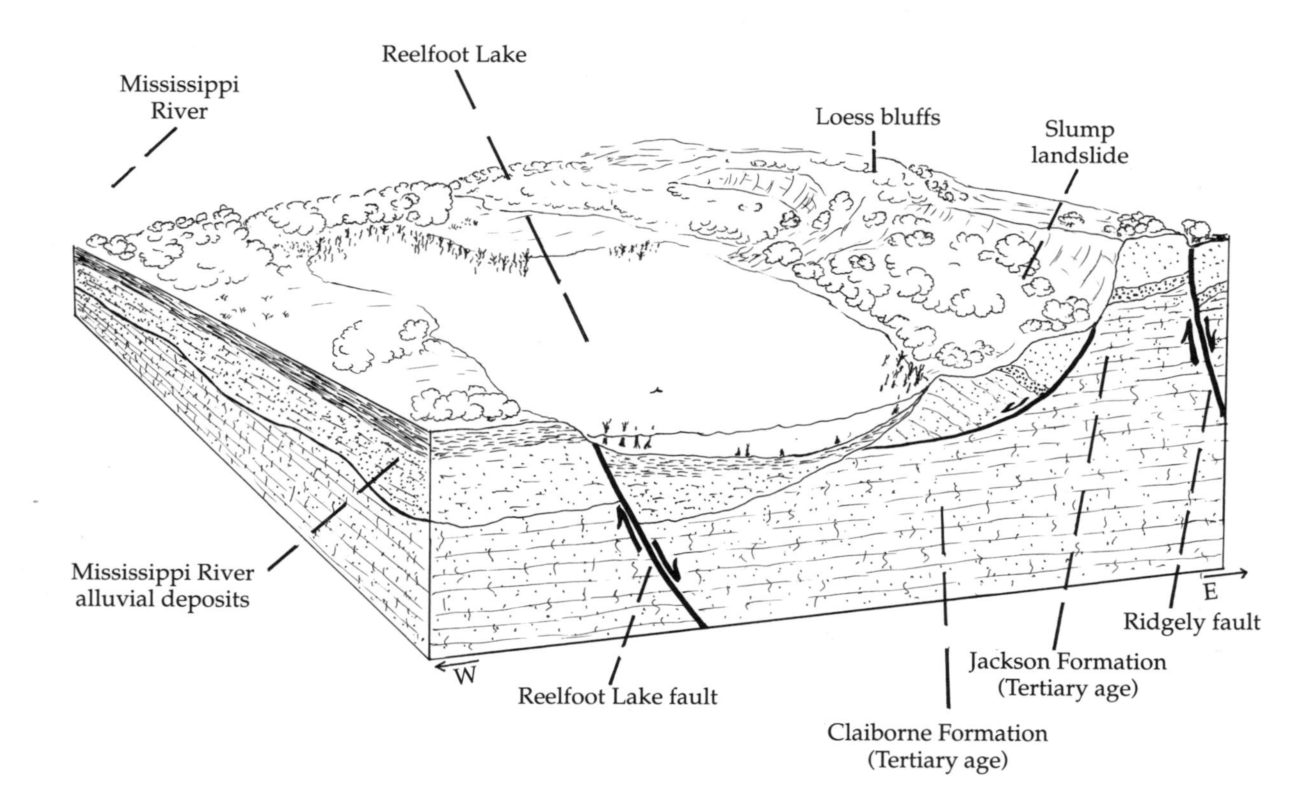

Fig. 136. This schematic block diagram illustrates the subsurface conditions in the Reelfoot Lake area that resulted from the devastating earthquakes of 1811 and 1812.

Drive around Reelfoot Lake (optional)

The 40-mile scenic drive takes you around the entire border of Reelfoot Lake, where you will see not only the lake but the backwaters environment, including a stop at the Airpark Inn. During the winter months the Airpark Inn area is a favorite place for viewing the bald eagles. The modern inn, operated by the state park, provides lodging and restaurant services.

The scenic drive begins at Reelfoot Lake State Park Visitors Center, open daily from 8:00 A.M. to 4:30 P.M., which has interpretive exhibits on the formation, history, and natural diversity of Reelfoot Lake and projections about its future. In addition, there is the boardwalk (described above) winding its way through the cypress trees out over the lake. At Log Mile 0.8 the drive passes by the Blue Bank Day Use Area, where there are lakefront picnic sites with tables and grills as well as places from which to fish.

The route passes over the Reelfoot Spillway at Log Mile 2.4 (fishing and picnic facilities are available here, too). This structure helps control the level of the lake by allowing the waters when at higher levels to flow out of the basin. In 1937 dynamite was found at the spillway—an effort by some to destroy the lake. A wet fuse saved the day. Controversy persists even today over who should control the lake level.

As you cross the spillway, you pass from Lake County into Obion County. At Log Mile 2.9 the route goes by the Reelfoot Lake State Park Campground. This scenic area, with lakeside campsites nestled among the cypress trees, has excellent views of the lake, especially at sunset.

At Log Mile 3.3, turn left, continuing on S.R. 22 (S.R. 21 veers off to the right). The ridge to the right of the roadway, which is the escarpment of the Gulf Coastal Plain province, is underlain by the yellowish-gray to gray claystone of the Jackson Formation (Tertiary age).

You encounter the small community of Samburg at Log Mile 6.1. Samburg, one of the oldest communities on Reelfoot Lake, is a good place to view bald eagles in the winter and wildflowers in the summer. In addition, you can find restaurants, motels, and shops here.

At Log Mile 7.0 you begin a drive along the base of the bluffs (the escarpment bluffs are on your right as you travel north on S.R. 22). The Jackson Formation and loess deposits are found in these bluffs but are not visible from the road at this point (you will be able to see them later). The Jackson Formation is composed of claystone and sand lenses (layers with a short lateral extent) containing thin lenses of silica and lignite. Just above the Jackson Formation is a gravel deposit overlain by the Pleistocene loess, a yellowish-brown, angular silt.

Near Log Mile 8.3 is the Kirby Pocket community. Notice the many cypress trees in the lake. Although the lake is being silted in by erosion of the bluff and upland area, the cypresses survive in the shallower water. At Log Mile 8.3 is the Kirby Pocket Visitor Center, open most weekends, which contains exhibits about the history and aquatic world of Reelfoot Lake.

On the right side of the road at Log Mile 10.1 is an exposure of the Jackson Formation in the roadway cutslope (Fig. 137). These horizontal rock units, consisting of a grayish-buff claystone, contain numerous plant fragments as well as lenses of gravel and sand. There are also many seams of whitish silica in the claystone at this location. At Log Mile 12.5 you leave S.R. 22 and turn left onto S.R. 157, which skirts the edge of the exposed Jackson Formation (to the right of the roadway).

Fig. 137. Along the scenic drive around Reelfoot Lake exposures of the Tertiary-age Jackson Formation can be seen near Log Mile 10.1. Numerous plant fragments can be found in the strata.

Continuing on S.R. 157, you find the Reelfoot Lake National Wildlife Refuge Visitors Center off to your left at Log Mile 13.5. The visitors center, open Monday through Friday from 8:00 A.M. to 4:00 P.M., provides information about the great diversity of fish, reptiles, mammals, birds, and plants found at Reelfoot Lake. There is also a museum with excellent exhibits on the wildlife. In addition, maps of the roads and trails in the wildlife refuge can be obtained there.

The Reelfoot Lake National Wildlife Refuge is one of three wildlife management areas at Reelfoot Lake (the others are the Reelfoot Lake State Park and Wildlife Management Area and Lake Isom National Wildlife Refuge, the latter about 2 miles south of the Reelfoot Lake Spillway). Altogether, the Reelfoot area contains a vast acreage of wetlands set aside for the protection of wildlife. These areas are particularly good for observing waterfowl.

At about Log Mile 14.5 is the turnoff to Grassy Island Wildlife Drive (near Walnut Log) on your left. You can drive or walk through the Grassy Island portion of the Reelfoot Lake National Wildlife Refuge. This is a fine area for observing woodland birds and mammals. The drive extends to the edge of Reelfoot Lake, where there is a boardwalk and observation platform, a beautiful location for watching the sunset over the cypress-filled waters (the round-trip from S.R. 157 is approximately 5 miles).

At Log Mile 15.5 the road climbs upon the bluff. Here you can see numerous exposures of the Jackson Formation and loess deposits. Researchers have found indications that this section of the bluff has experienced several slump-type landslides: for example, these strata have slid down at least 50 feet to the northwest and, though once flat-lying, they are now dipping from 23° to 44° to the southeast. Although these seem to be features of a slump block, it is possible that they extend beneath the surface and may be normal faults (Blythe, McCutchen, and Stearns 1975: 74).

Near Log Mile 17.0 you leave the bluff area and begin traversing the flat floodplain of the Mississippi River. From this point to the end of the route at the park visitors center, you are traveling over alluvial floodplain deposits of the Mississippi River. Notice that the topography is flat and poorly drained.

At Log Mile 17.8, S.R. 157 ends and Ky. 311 begins. After traveling about 1 mile into Kentucky, turn left at the crossroads (Log Mile 18.8) onto Road 1282. Along Road 1282 is the Long Point portion of the National Wildlife Refuge. (If you are interested in a drive through open fields and woods to view wildlife, you may like to take a detour and investigate one of the several gravel roads that turn left off Road 1282 and lead into the refuge. This is a good area for observing migratory birds in the spring and fall and thousands of ducks and Canada geese during the winter.)

Continue on the paved Road 1282 until the intersection with Ky. 94 at Log Mile 22.4. There you turn left (west) onto Ky. 94. It is now 1.4 miles to the Tennessee state line. At Log Mile 23.8 the route reenters Tennessee. Ky. 94 now becomes Tennessee S.R. 78 in Lake County. The Mississippi River is to the right of the road, but you cannot see it from the road.

The route intersects S.R. 213 at Log Mile 26.9. Turn left and follow S.R. 213 to the Airpark Inn (about 3 miles). At the Airpark Inn, located on the shore of Reelfoot Lake, there are orientation talks about observing the bald eagles. The inn and restaurant are built out over a small, shallow part of Reelfoot Lake. An observation pier extending out into the lake from the inn provides good opportunities for seeing birds at almost any season (and bald eagles during the winter). This is a lovely place to stay overnight, but you probably need to call or write Reelfoot Lake State Park ahead for reservations (see the additional information at the end of this section).

Note: During the winter months there are regularly scheduled bus tours from the Airpark Inn to look for bald eagles and waterfowl. These tours, conducted daily from December 1 to mid-March by state park personnel, generally begin at 10:00 A.M. and last about 2 hours. There is a moderate fee, and registration is required. For information, call or write the park.

Retrace the route back to S.R. 78 from the Airpark Inn; turn left (south) onto S.R. 78 (Log Mile 32.9). To your left at Log Mile 37.5 is a road that leads to the south entrance of the Reelfoot Lake Wildlife Management Area. (This detour provides a fine opportunity for a drive and walk to look for wildlife in the area. When you reach the end of the road, you will find a walking trail that follows an old dirt roadbed to Donaldson Ditch, which connects two major portions of Reelfoot Lake. The foot trail is approximately 2 miles round-trip.)

At Log Mile 38.5 the route intersects S.R. 212 on the right. Continuing on S.R. 78, you enter the Tiptonville city limits at Log Mile 40.0. At Log Mile 40.4 is the junction with S.R. 22. Here you end the circular route around Reelfoot Lake and your side trip. Return to I-40 at Jackson, via S.R. 78 and U.S. 412.

Additional information (food, lodging, schedules of events):

Superintendent's Office
Reelfoot Lake State Resort Park
Route 1
Tiptonville, TN 38079
(901) 253-7756

Crockett County Chamber of Commerce
320 South Bells Street
Alamo, TN 38001
(901) 696-5120

Dyersburg–Dyer County Chamber of Commerce
2455 Lake Road
P.O. Box 906
Dyersburg, TN 38025-0906
(901) 285-3433

Jackson Area Chamber of Commerce
197 Auditorium Street
P.O. Box 1904
Jackson, TN 38301
(901) 423-2200

Obion County Chamber of Commerce
215 South First Street
P.O. Box 70
Union City, TN 38261
(901) 885-0211

Reelfoot Area Chamber of Commerce
1116 Parnell
Tiptonville, TN 38079
(901) 253-8144

9. Fort Pillow State Historic Area

Destination: Chickasaw Bluffs, Fort Pillow State Historic Area
Route: U.S. 412, S.R. 88, U.S. 51, S.R. 87, S.R. 207; Park Service road; begins at I-40, Exit 79 (Map 22; alternate route from Exit 65 via Brownsville given below)
Cities and towns: Alamo, Ripley, Henning
Trip length: 71.8 miles
Nature of the roads: four-lane and two-lane, paved highway, some flat, straight sections; many curvy, hilly portions, especially along S.R. 87
Beginning elevation: ±350 feet
Ending elevation: ±370 feet
Special features: Rolling landscape of Gulf Coastal Plain province; loess deposits and Chickasaw Bluffs; Civil War fortifications; Mississippi River
Hiking trail (optional): Chickasaw Bluff Civil War Fortifications Trail
U.S.G.S. 7 1/2 minute quadrangle maps: Gilt Edge 407 SE, Gold Dust 407 NE
For reference: Gulf Coastal Plain, Mississippi River, loess, Jackson Formation

Directions. From I-40 take Exit 79 and turn right (north) onto U.S. 412, which you follow for 18.2 miles from Madison County into Crockett County. In Crockett County turn left onto S.R. 88 and follow it to S.R. 209 in the town of Gates (you will pass through the town of Maury City). Turn left onto S.R. 209 and follow it for 0.2 mile; then turn right onto S.R. 180. Follow S.R. 180 for 2.5 miles to U.S. 51. Turn left (south) onto U.S. 51 and follow it for approximately 14 miles to the junction with S.R. 87; there you turn right onto S.R. 87 and follow it for approximately 17 miles (you pass by Fort Pillow State Prison). Turn right from S.R. 87 onto S.R. 207, which you follow for about 1 mile to the entrance to Fort Pillow State Historic Area. To reach the interpretive center and trailhead, follow the park road for approximately 2 miles.

Alternate route. From I-40 take Exit 65 and turn onto U.S. 70W, which you follow to Brownsville. In Brownsville turn onto S.R. 54 and then shortly turn right onto S.R. 19. Follow S.R. 19 for about 1 mile, turning left onto S.R. 87. Take S.R. 87 to its intersection with U.S. 51 (you pass

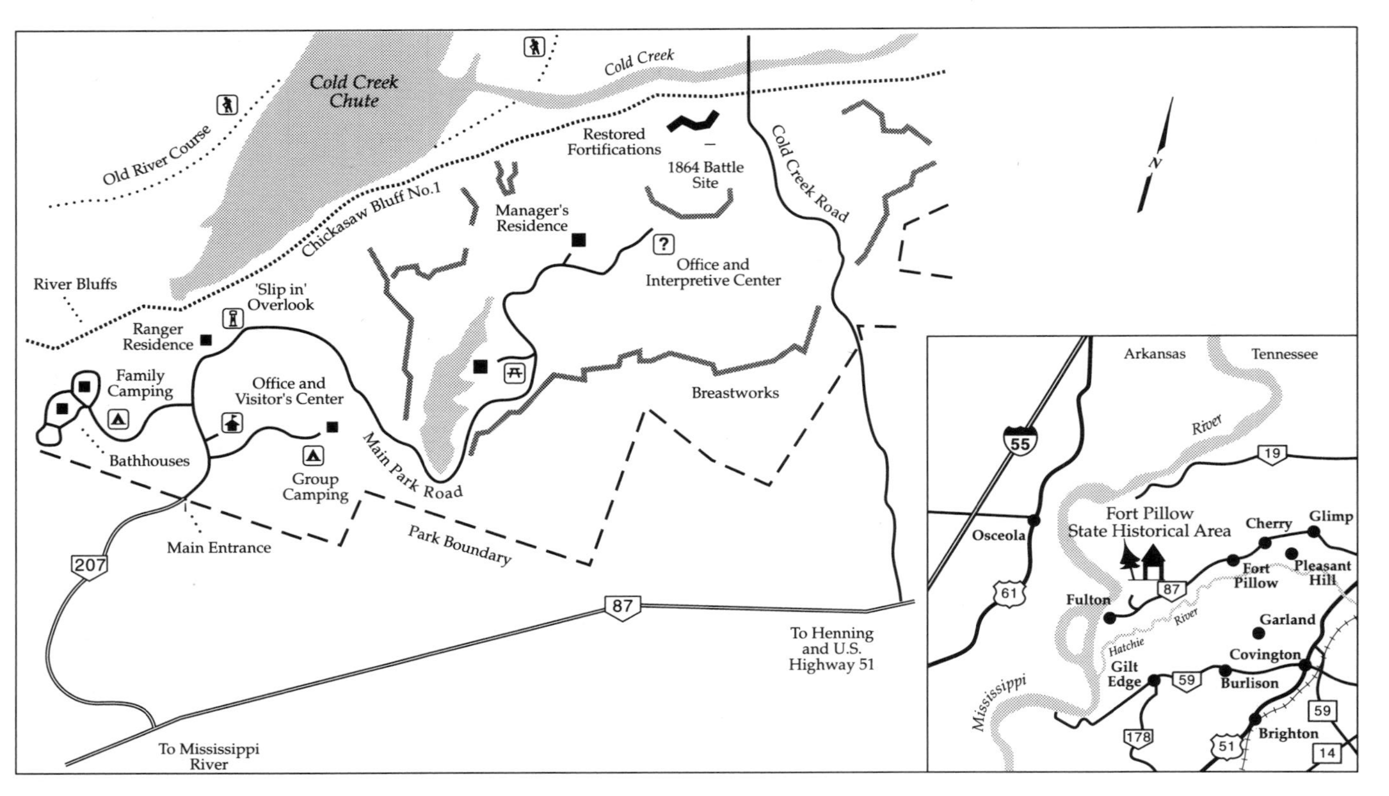

Map 22. Fort Pillow State Historic Area. Courtesy of Tennessee Department Conservation and Environment, State Parks Division.

through the town of Henning). At the intersection of S.R. 87 and U.S. 51 (Log Mile 50.9), the alternate route intersects the main route given in the directions above. At this point both the main route and the alternate route follow S.R. 87 past U.S. 51 toward the Mississippi River for approximately 17 miles, passing Fort Pillow State Prison. Turn right off S.R. 87 onto S.R. 207, which you follow for about 1 mile to the entrance to Fort Pillow State Historic Area.

Route details. Taking this route, you will travel to the Mississippi River and view the restored battlements of Fort Pillow, a Civil War battle site. Along the way you will see the rolling landscape of the Gulf Coastal Plain province of West Tennessee and the bluff-forming character of the windblown loess deposits.

As you begin this side trip, the route quickly enters a broad floodplain of the South Fork of the Forked Deer River at Log Mile 2.0. The floodplain area (to the left of the highway) is being used for agriculture. At Log Mile 5.5 you leave the floodplain area and again begin traversing the loess-covered landscape of the Gulf Coastal Plain. For most of this route you will see cultivated fields where cotton, soybeans, wheat, corn, and milo are growing. This is the heart of the agriculture belt in Tennessee.

The route passes through three counties: Madison (Log Miles 0.0–7.5), Crockett (Log Miles 7.5–30.5), and Lauderdale (Log Miles 30.5–71.8). For most of the way, the geology lies buried beneath the loess deposits. At your destination, however, you will get to see the effects of a meandering Mississippi River, the bluff-forming characteristics of the loess deposits of the Quaternary period, and exposures of the Jackson Formation of the Tertiary period.

As you travel across Crockett County on S.R. 88, the route passes through the community of Frog Jump at Log Mile 29.3 and again encounters the floodplain of the South Fork of the Forked Deer River at Log Mile 30.5. The mile-wide floodplain is located about 2 miles east of Gates along the boundary between Crockett and Lauderdale counties.

In Lauderdale County, S.R. 88 intersects S.R. 209 at Log Mile 34.3 in the community of Gates. The route then follows S.R. 209 for only 0.2 mile, soon turning right onto S.R. 180. You then follow S.R. 180 across a hilly section of landscape for 2.5 miles until the route intersects U.S. 51 at Log Mile 37.0. There you turn left onto U.S. 51 and follow it for about 14 miles. As you cross the drainage divide between the South Fork of the Forked Deer River and the Hatchie River, the landscape becomes pronouncedly eroded.

At Log Mile 50.9, on U.S. 51, the route turns right on S.R. 87 at Henning, the hometown of the late Alex Haley, author of *Roots*. The two-lane S.R. 87 becomes curvy as it traverses the crest of a narrow ridge from Log Mile 50.9 to Log Mile 59.1. This rolling topography is underlain by the Jackson Formation and capped by silty, buff-brown loess deposits.

At Log Mile 61.2 the route takes you by the main buildings of the Fort Pillow State Prison Farm. Here state prisoners operate a row crop farm, producing cotton and soybeans as well as vegetables for the prison. Although the area is secure and heavily guarded, it is particularly inadvisable to pick up hitchhikers along here!

Along S.R. 87 (Log Mile 67.0–68.3), you can see portions of the Lower Hatchie National Wildlife Refuge. The refuge continues along the south side of S.R. 87, but the side trip route turns off S.R. 87 at Log Mile 68.3 onto S.R. 207. Numerous wetland birds and other animals can be observed here. (Further information can be obtained at the refuge office, located 3 miles south of Brownsville at the intersection of I-40 and S.R. 76.)

The route intersects S.R. 207 at Log Mile 68.3. Turning right onto S.R. 207, you immediately see the effects of the loess deposits as you begin driving up a large hill underlain by this material. The deeply incised gullies you notice along the sides of the hill are a result of the erosive character of the loess soil. When on a nearly vertical exposure, the loess is very resistant to erosion because the angularity of the grains of silt produces a steep angle of repose (near vertical). When exposed on flatter slopes, however, the loess becomes unstable and is subject to the erosive powers of running water.

After traveling 1.1 miles on S.R. 207, you enter Fort Pillow State Historic Area at Log Mile 69.4. The park office and visitors center are located just beyond the entrance to the park. The trailhead to the restored fortifications and the interpretive center are at the end of the main park road, 2.5 miles from the park entrance (Log Mile 71.9).

Hiking Trail to Chickasaw Bluffs and Restored Civil War Fortifications (optional)

Location: Trail begins at north end of parking lot at interpretive center
Start: Interpretive center parking area, elevation ±370 feet
End: Restored fortifications atop Chickasaw Bluffs, elevation ± 360 feet; Mississippi River, elevation 215 feet
Nature of the trail: Gravel and bare soil foot trail, moderate difficulty; hilly, with many steps and one swinging bridge
Distance: 1.0 mile, round-trip
Special features: Chickasaw Bluffs, contact between loess deposits and Jackson Formation (Tertiary period), natural landslides, restored Civil War fortifications

This hiking trail takes you into a historic area with a geologically interesting environment, where you can see the bluff-forming loess deposits and the results of past landslides (Fig. 138). At the edge of the Chickasaw Bluffs, you enter the restored Civil War fortifications of Fort Pillow (Fig. 139).

Caution: Although several footpaths lead down over the bluffs from the main trail, you are advised not take any of them because they are dangerous. Groundwater seeps along the contact between the overlying loess deposits and the Jackson Formation (near the base of the bluffs) produces very slick surfaces, at times impossible to climb back up.

Along the trail to the Chickasaw Bluffs, you will encounter the escarpment between the Gulf Coastal Plain deposits and the Mississippi River Alluvial Floodplain.

Historical significance. Fort Pillow was named after General Gideon J. Pillow, a hero of the Mexican War. The fort was originally constructed by Confederate forces to provide defense against a Union naval attack from the Mississippi River, which at that time was located at the base of the bluffs. In the summer of 1862 the Union boats attacked Fort Pillow, but without much success. The Confederate forces, however, decided to leave the fort in order to remain in contact with their other armies in the region, and then the Union forces occupied it for nearly two years. In April 1864, General Nathan Bedford Forrest led about 1,500 Confederate soldiers in an attack on Fort Pillow. After General Forrest failed in his attempts to persuade the Union forces to surrender,

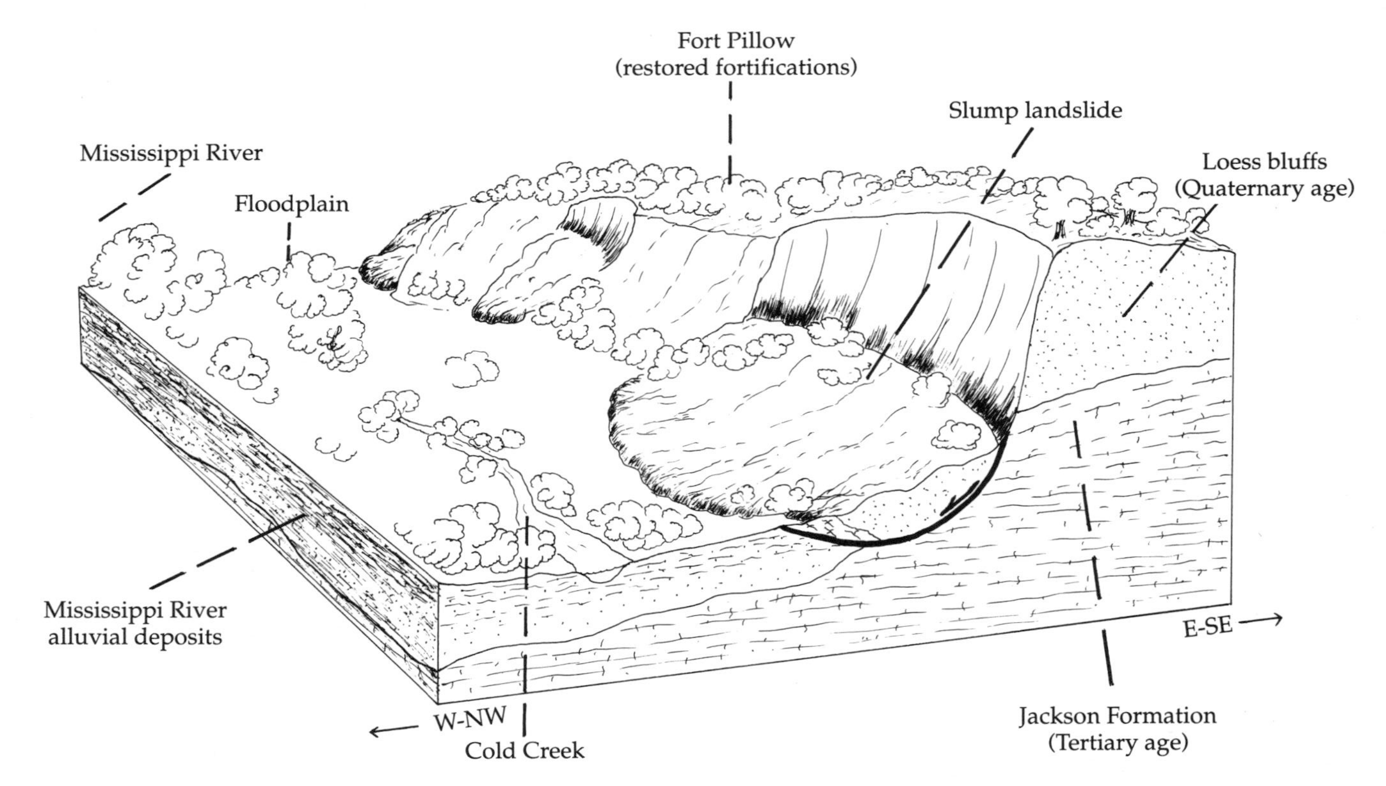

Fig. 138. Fort Pillow State Historic area is located on top of a loess bluff overlooking the Mississippi River. Large slump landslides have occurred along the bluff face, producing half-circle–shaped scars in the bluff surface.

Fig. 139. Fort Pillow State Historic Area provides a museum and interpretive center of the historic aspects of the Civil War Battle of Fort Pillow.

the Confederates stormed the fort. Because almost half of the 550 Union soldiers at the fort were black and the Union soldiers suffered high casualties, historians have found this battle most controversial. After April 12, 1864, Fort Pillow was unoccupied (Mainfort 1980: 1–4).

Trail description. You begin the trail crossing old Civil War earthen breastworks (not restored) and gradually descending the hill. After about 150 yards you encounter a long series of steps that take you to near the bottom of a bluff. Here you can see several ancient landslide scars to the left of the trail, just before you cross the swinging bridge (Fig. 140).

To the left of the trail, you see soil banks with a half-circle pattern. These are old landslide scarps—the "heads" of the landslides, where the land surface has cracked and settled downslope. Such landslides are a natural type of mass wasting along the Chickasaw Bluffs adjacent to the Mississippi River (Fig. 141).

Groundwater percolating down through the silty loess material encounters a natural barrier where the loess comes in contact with the Jackson Formation. The Jackson Formation consists of light-brownish-gray sand and clay layers, the latter impermeable. Because the downward

Fig. 140. On the trail to the restored battlements of Fort Pillow you will cross a suspended footbridge that spans a gully eroded into the loess soils.

Fig. 141. Along the foot trail to the restored Fort Pillow you will be able to see effects of weathering on the loess soils. Here, near the suspended bridge, you can see slump landslides where brown loess soils have slid over the grayish clays of the Jackson Formation.

migration of the groundwater is impeded at the top of the Jackson Formation, there is an increase of seepage pressures. The groundwater, seeking areas of less resistance, seeps out at the surface, near the base of the loess deposits.

The increase of groundwater near the base of the loess material also increases the water pressure between the grains of silt and causes the loess material to have a loss of shear strength (the ability of the soil to withstand failure from an overburden of weight). A landslide results when the soil shear strength drops below the strength required to maintain stability. Also contributing to the landslides along the Chickasaw Bluffs are the vibration effects from earthquakes along the nearby New Madrid fault zone.

As you cross the swinging bridge, look down into the creek channel below. The gray clay and sand materials in the lowermost portions of the gully are exposures of the Jackson Formation (Tertiary period). The light brown, silty sand that comprises the upper two-thirds of the gully area are loess deposits (Quaternary period).

After crossing the bridge, continue up the hill to the left. At the top of the hill (about 200 yards) are the restored fortifications, which are located near the edge of the Chickasaw Bluffs (Fig. 142). At the base of the bluffs is the old channel of the Mississippi River. It is now an oxbow lake, Cold Creek Chute Lake, the result of a changing river course. The numerous semicircular earthen banks that you see along the edge of the bluff near the restored fortifications are slump scars, the results of landslides (like the ones near the swinging bridge).

The upper two-thirds of the bluff is composed of loess material, the angular silt deposited by strong westerly winds along the eastern banks of the Mississippi River. Because the loess material has a very steep angle of repose (nearly vertical), it is more stable on steep slopes (45° to vertical) and less stable—more subject to erosion—when the land surface is flatter (slopes of 15° to 45°). As a result, the steep slopes of the bluffs are somewhat stable, though occasionally subject to slumping (landsliding) due to erosion and possibly vibration effects from earthquakes along the nearby New Madrid fault (see Side Trip 8 to Reelfoot Lake).

Fig. 142. The restored battlements of Fort Pillow are located on a loess bluff. The fort overlooked a bend of the Mississippi River during the Civil War, but now, due to the meandering nature of the river, the Mississippi is over a mile away from the bluff.

The fortifications, including the cannons, are restored to scale. Interpretive signs give information about the battle of Fort Pillow, April 12, 1864. A very small scale model of the Fort Pillow battle is on display at the visitors center. The hand-crafted model, only 2 feet by 4 feet in size, has men and horses smaller than the diameter of a dime (Sullivan 1990: 25–26).

To return to the parking area, retrace the trail. If you would like a longer hike, the Bluff Trail continues as the Historical Trail route. Ask at the park office for hiking information.

Additional information (food, lodging, schedules of events):

Fort Pillow State Historic Area
Route 2, Box 109
Henning, TN 38041
(901) 738-5581

Lower Hatchie National Wildlife Refuge
P.O. Box 187
Brownsville, TN 38012
(901) 772-0501

Brownsville–Haywood County Chamber of Commerce
108 North Lafayette Avenue
Brownsville, TN 38012
(901) 772-2193

Crockett County Chamber of Commerce
320 South Bells Street
Alamo, TN 38001
(901) 696-5120

Jackson Area Chamber of Commerce
197 Auditorium Street
P.O. Box 1904
Jackson, TN 38301
(901) 423-2200

Lauderdale County Chamber of Commerce
110 South Jefferson Street
Ripley, TN 38063
(901) 635-9541

Appendix 1. Caving Clubs

East Tennessee

Appalachian Grotto
385 South Bellwood Road
Morristown, TN 37813

Chattanooga Valley Grotto
P.O. Box 16762
Chattanooga, TN 37416-0762

East Tennessee Grotto
320 Wardly Road
Knoxville, TN 37922

Holston Valley Grotto
1504 Chickees Street
Johnson City, TN 37604

Mountain Empire Grotto
P.O. Box 176
Blountville, TN 37617

Smoky Mountain Grotto
Box 8297, UT Station
Knoxville, TN 37996

Middle Tennessee

Cumberland Valley Grotto
Route 2, Box 145-A
Liberty, TN 37095

Nashville Grotto
P.O. Box 23114
Nashville, TN 37202

Tennessee Central Basin Grotto
2319 Reidhurst Drive
Murfreesboro, TN 37130-9806

West Tennessee

West Tennessee Grotto
581 Winford Drive
Collierville, TN 38017-1003

Appendix 2. Rock and Mineral Clubs

East Tennessee

Chattanooga Geology Club
4117 Ealy Road
Chattanooga, TN 37412

Knoxville Gem and Mineral Society
5604 Malmsbury Road
Knoxville, TN 37921

Smoky Mountain Mineral Society
4200 A Bennett Branch Road
Hartford, TN 37753

Tennessee Eastman Rock and Mineral Club
P.O. Box 511
Kingsport, TN 37662

Tennessee Valley Rock and Mineral Club
2529 Allegheny Drive
Chattanooga, TN 37421

Unaka Rock and Mineral Society
416 Highland Avenue
Johnson City, TN 37604

Middle Tennessee

Middle Tennessee Gem and Mineral Society
P.O. Box 1256
Murfreesboro, TN 37133-1256

Middle Tennessee Rockhounds
109 Rosebank Avenue
Nashville, TN 37206

West Tennessee

Memphis Archaeological and Geological Society, Inc.
2959 Sky Way Drive
Memphis, TN 38127

Appendix 3. Mineral Use and Laws of Geology

Important Mineral Uses

Apatite - Fertilizer
Asbestos - Insulation
Barite - Paint pigment
Bauxite - Aluminium ore
Calcite - Cement, flux
Chalcopyrite - Copper ore
Corundum - Abrasive, gems
Dolmite - Building stone
Feldspar - Flux, glaze constituent
Florite - Flux, glass
Galena - Lead ore
Garnet - Gems
Gypsum - Portland cement
Halite - Salt, soap
Hematite - Iron ore
Limonite - Iron ore
Magnetite - Iron ore
Siderite - Iron ore
Kaolin - Ceramics
Mica - Electrical industry
Olivine - Refractory mineral
Pyrite - Sulphuric acid
Quartz - Gems, abrasive
Sphalerite - Zinc ore
Talc - Soapstone

Some Common Laws of Stratigraphic and Historical Geology

Law of Uniformitarianism. The key is the present to the past. This law, probably the most fundamental law of the geological science, states that all of the earth's features (landforms, rocks, etc.) of the past and the present have been formed by the same processes now active on and within the earth. The laws of nature have not changed during the billions of years of earth's history, and hence an understanding of the processes at work on the earth today is essential to a proper understanding of the geologic past.

Law of Original Horizontality of Sedimentary Rocks. Sedimentary layers are deposited essentially parallel to the depositional surface, and since the latter is essentially horizontal, the resulting sedimentary formation that is not flat-lying has been deformed, i.e., tilted, folded or faulted, subsequent to deposition.

Law of Superposition. In a sequence of sedimentary rock formations (and lava flows) the lowest or bottom formation is the oldest and the succeeding overlying layers are successively younger. This law does not prevail when the layers have been overturned.

Law of Structural Relationship. A deformation (such as folding or faulting) is younger than (or occurred after) the youngest rocks deformed and older than the oldest rocks not deformed.

Law of Igneous Relationship. An igneous intrusion or extrusion is younger than the youngest rocks that it intrudes or effects in any way (i.e., arches or metamorphoses) and older than the oldest rocks that it does not intrude or effect in any way.

Law of Metamorphic Relationship. When a series of metamorphic rocks are unconformably overlain by a series of unmetamorphosed rocks, the metamorphism occurred after (or is younger than) the youngest rocks metamorphosed and before (or is older than) the oldest rocks not metamorphosed.

Law of Faunal Succession. Each formation or layer has its own distinctive fauna (assemblage of fossil animals) and/or flora (assemblage of plants) unlike that of the layers above and below. Furthermore, the overlying or younger layers (unless they have been overturned) contain more advanced types of fossil animals and/or plants than the lower or older ones.

(Taken from Klepser, 1967)

Glossary

Abrasion. The mechanical wearing of solid materials by impact and friction

Acid mine drainage. Acidic ground and / or surface water emitted from a mine site, usually containing a percentage of sulfuric acid

Age. Any great period of time in the history of the earth or the material universe marked by special phases of physical conditions or organic development

Alluvium. Sediments formed by rivers and streams

Angular unconformity. An arrangement in which older, deformed, stratified rocks have been truncated by erosion and younger layers have been laid down upon them with a different angle of inclination

Anomaly. A deviation from uniformity; a local feature distinguishable by a technical measurement over a larger area

Anticline. A fold or arch of stratified rocks in which the strata dip in opposite directions from a common ride or axis

Aquifer. A porous, permeable rock layer from which water may be obtained

Argillaceous. Composed of clay, or containing clay minerals

Argillite. Dark, fine-grained rock without cleavage or schistosity; resulting from low-grade metamorphosis of claystone or mudstone

Arkose. A sandstone containing at least 25 percent feldspar, usually derived from erosion of granitic rocks

Asphalt. A brown to black solid or semisolid bituminous substance, occurring in nature but also obtained as the residue from the refining of certain petroleums and then known as artificial asphalt

Asthenosphere. The shell of weakness within the earth, some tens of kilometers below the surface, where plastic movements take place to permit isostatic adjustments

Attitude. The relation of some directional feature in a rock to a horizontal plane

Auger drilling. A drilling process, usually in soil, in which the cuttings are continuously removed mechanically from the bottom of the bore without the use of fluids

Avalanche. A large mass of snow or ice, sometimes accompanied by other material, moving rapidly down a mountain slope; sometimes applied to rapidly moving landslide debris

Axis. A straight line, real or imaginary, about which a body rotates or may be imagined to rotate

Basalt. A fine-grained, dark, igneous rock composed of plagioclase feldspar and dark, heavy minerals (predominantly pyroxenes); usually extrusive as lavas

Basement. An older rock mass, usually igneous or metamorphic, on which younger rocks have been deposited

Bedding. Layering in sedimentary rocks

Bedding plane fault. A fault surface which is parallel to the bedding plane of the constituent rocks

Bedrock. Solid rock underlying weathered or transported material

Bench. A level or gently sloping area interrupting an otherwise steep slope; commonly seen along highway cutslopes

Block diagram. A three-dimensional perspective representation of geologic or topographic features, showing a surface area and generally two vertical cross-sections

Block field. An accumulation of rock talus that is wider than it is long; usually found on mountain slopes below sources such as resistant rock bluffs or ledges

Block stream. An accumulation of rock fragments (usually boulder size and larger) that is elongated in the slope direction; lengths can range from 30 feet to over 3,000 feet

Boulder. A large, rounded block of stone lying on the surface of the ground or sometimes embedded in loose soil

Brachiopod. A marine shellfish with two bilaterally symmetrical shells

Brush barrier. A mound or pile of cut trees, limbs, and undergrowth used to reduce siltation from erosion of excavated areas

Buttress. A structure of masonry, wood, or quarried rock that gives support or stability to a slope

Calcareous. Containing calcium carbonate

Calcite. A mineral, calcium carbonate ($CaCO_3$), the principal constituent of limestone

Cambrian. The first period of the Paleozoic Era, from 500 to 600 million years ago

Carbonaceous. Containing or composed of carbon; pertaining to sediment containing organic matter

Carbonate. Containing carbon and oxygen in combination with sodium, calcium, or other elements, particularly in limestone or dolomite

Carboniferous. The interval of time representing the Mississippian and Pennsylvanian periods, from 270 to 350 million years ago

Cave. A natural cavity, recess, chamber, or series of chambers beneath the surface of the earth, generally produced by the solution of limestone

Cedar glade. A local vegetation dominated by cedar trees, usually found on shallow limestone soils

Cenozoic. The most recent of the four eras into which geologic time is divided, extending from the close of the Mesozoic Era to and including the present; it includes the periods called Tertiary and Quaternary in the United States

Chalcopyrite. A widely occurring ore of copper which also contains iron and a sulfur compound known as fool's gold

Check dam. A small mound of rock or other debris placed in the channel of a small stream or drainage ditch to impede the flow of water and filter out sediment

Chert. Also called flint, a hard, extremely dense, sedimentary rock consisting predominantly of cryptocrystalline silica; a variety of the mineral quartz

Clastic. Composed of fragmental material derived from preexisting rocks

Cleavage. The tendency for rocks to split along definite planes which generally have no relation to bedding

Cohesion. The resistance of a material, rock, or sediment against shear along a surface which is under pressure

Colluvium. Loose and incoherent deposits, usually at the foot of a slope or cliff and brought there chiefly by gravity

Competent. Capable of sustaining stress without being greatly deformed

Compressible soils. Soils that can change volume due to the expulsion of only the pore water when subjected to load

Conformable. Deposited without significant disturbance or removal of previously deposited strata

Conglomerate. Rock composed of rounded, waterworn fragments of older rock, usually in combination with sand

Continental rocks. Rocks making up the continents, lighter in color and density than those formed in ocean basins

Continental shelf. That part of the continental margin between the shoreline and the continental slope that is gently sloping to a depth of about 600 feet below sea level

Convection. A process of mass movement of portions of any fluid medium (liquid or gas) in a gravitational field as a consequence of different temperatures in the medium and hence different densities

Corridor. A certain path or route location on which a study is centered

Craton. A relatively immobile part of the earth's crust, generally of large size, composing the stable portion of most continents

Cross-bedded. Deposited at an angle to the horizontal; also called current bedding

Cross-section. A profile portraying an interpretation of a vertical section of the earth explored by geological and/or geophysical methods; a short profile section usually perpendicular in orientation to a profile or survey line

Crust. The outermost layer or shell of the earth

Crystalline. Composed wholly of crystalline mineral grains; that is, igneous and metamorphic rocks as distinct from sedimentary rocks

Cut. An excavation of soil and rock in a hill, ridge, or mountain, where the result is usually an artificially smooth or sculptured slope in the earth

Debris. Rock and mineral fragments produced by weathering of rocks; synonymous with detritus

Debris flow. A general designation for all types of rapid flowage involving debris of various kinds and conditions

Debris slide. The rapid downward movement of predominantly unconsolidated and incoherent earth and debris in which the mass does not show backward rotation but slides or rolls forward, forming an irregular deposit

Deformation. Change of shape or attitude of a rock body by folding, shearing, fracturing, compression, etc.

Devonian. The fourth period of the Paleozoic Era, from 350 to 400 million years ago

Diorite. A coarse-grained igneous rock, heavier and darker than granite, composed of plagioclase feldspar, amphibole, pyroxene, and minor quartz

Dip. The angle at which a stratum or any planar feature is inclined from the horizontal

Discontinuity. A sudden change in the character of rock, such as joints, cleavage, bedding, faults, or contacts between similar or dissimilar materials

Dolomite. A rock composed essentially of the mineral dolomite, $(CaMg)CO_3$

Dolostone. A calcareous sedimentary rock composed of the mineral dolomite

Dome. In topography, a roughly circular and upwardly convex landform; a structural dome in sedimentary rocks involves an outward dip or inclination of the beds in all directions

Dye tracing. The use of a dye agent injected into the groundwater to distinguish the flow patterns of groundwater

Earth flow. A form of mass movement in which relatively unconsolidated surface material, deeply weathered, flows down a hillside

Ecosystem. Ecologic system; an organic community and its physical environment

Embankment. An artificial accumulation of soil and rock placed by mechanical means in order to build a roadway; used to fill across valleys

Embayment. An indentation along a shoreline, mountain front, or any other natural linear feature

Encapsulation. Enclosing entirely within a substance; enclosing acid-producing rock within a compacted soil or synthetic material to prevent leakage of air or contaminants

Erosion. The process of disintegration and removal of the rocks at the earth's surface by weathering and moving water, wind, ice, or landslide

Escarpment. A cliff or steep slope edging a region of higher land

Extrusive rocks. Igneous rocks derived from magmas or magmatic materials poured out or ejected at the earth's surface

Fault. A fracture in the earth's crust along which rock on one side has been displaced relative to rock on the other

Fault scarp. A cliff formed by a fault, usually modified by erosion

Feldspar. A group of abundant light-colored, rock-forming minerals belonging to the silicate class

Fill. The placement by mechanical means of soil and rock to form a mound or raised earth structure

Floodplain. The surface of relatively smooth land adjacent to a river channel, constructed by the present river and covered with water when the river overflows its banks at times of high water

Fluvial. Of or pertaining to rivers; produced by river action

Foliation. The parallel alignment of platy mineral grains or flattened aggregates in a metamorphosed or sheared rock

Fold. A curve or bend of the rock strata, usually the result of deformation

Foldbelt. A linear region that has been subject to folding and other deformation during mountain building

Footwall. The underlying side of a fault

Fossil. The remains or traces of a plant or animal, preserved in rock

Formation. A distinctive body of some one kind or related kinds of rocks, selected from a succession of strata for convenience in mapping, description, and reference

Gabbro. The coarse-grained equivalent of basalt

Geomorphology. A branch of geology that deals with the earth's surface features and landforms

Geosyncline. An elongate depositional basin of continental proportions which is filled by sedimentary rocks over a long period of geologic time

Gneiss. A visibly crystalline metamorphic rock possessing mineral layering or foliation but not easily split along foliation surfaces

Grade. Measure of inclination of a surface, such as a roadway or land surface, expressed in percent

Granite. A visibly grained igneous rock composed essentially of alkali feldspars and quartz; commonly, any rock of this composition and texture, whether igneous or metamorphic in origin

Granitic. Pertaining to granite, or similar to granite in composition or texture

Granitization. The metamorphic transformation of nongranitic rocks to granite-like rocks

Graywacke. A gray sandstone consisting of poorly sorted, angular grains of quartz and feldspar, as well as abundant rock fragments

Groundwater. Subsurface water filling rock pore spaces, cracks, or solution channels

Group. A stratigraphic unit consisting of several formations, usually originally a single formation subdivided by subsequent research

Hanging wall. The overlying side of a fault

Hematite. A common iron mineral (Fe_2O_3)

Hogback. A ridge composed of a resistant layer within steeply tilted, eroded strata

Hydrostatic pressure. Pressure caused by weight of water in water-bearing rock or soil layers

Igneous. A term applied to rocks formed by crystallization or solidification from natural silicate melts, generally at temperatures between 600° C and 1000° C

Index fossil. Guide fossil; a fossil characteristic of an assemblage of fossil organisms and so far as known restricted to it

Interglacial. The time between major advances of continental glaciers

Intrusion. An igneous rock mass formed by the process of emplacement of magma in preexisting rock

Intrusive. Having penetrated, while fluid, into or between other rocks but solidifying before reaching the surface

Invertebrates. Animals without backbones

Island arch. A curved chain of islands, generally convex toward the open ocean, margined by a deep submarine trench and enclosing a deep sea basin

Joint. A fracture in rock along which no appreciable movement has occurred

Kaolinite. A white to tan silicate clay mineral usually occurring in colloidal masses; as a secondary mineral resulting from the weathering of feldspar, also found as clay deposits in the soil

Karst. A distinctive type of landscape in which solution in limestone layers has caused abundant caves, sinkholes, and solution valleys, often with red soil residue

Lagoon. A shallow stretch of sea water adjacent to the sea and partly or completely separated from it by a low, narrow strip of land

Landfill. An area of land excavated in trenches, in which garbage is placed and then covered by the excavated soil

Leachate. The liquid material that results from the passage of groundwater or other substances through subject material such as mine spoils, garbage, or soil wastes

Limestone. A sedimentary rock composed of calcite, the mineral form of calcium carbonate; loosely, either limestone or dolomite, a very similar rock that includes considerable magnesium in its chemical composition

Liquefaction. The process of liquefying or the state of being liquid; saturated silts liquefied by vibration

Lithology. The description of rocks on the basis of such characteristics as color, mineralogic composition, structures, and grain size

Lithosphere. The earth's crust; the outermost portion or shell of the globe, as distinguished from the underlying barysphere or centrosphere

Loess. Fine-grained silt, produced from glacial outwash and redeposited by wind

Macroscopic. Observations made with the unaided eye, as opposed to microscopic, made with the aid of a microscope

Magma. A hot, mobile silicate mixture of crystals and melt within the earth's crust

Mantle. The zone of the earth below the crust and above the core (to a depth of 3,480 meters)

Marble. Recrystallized limestone or dolomite; a metamorphic rock

Marl. A calcareous clay, or mixture of clay and particles of calcite or dolomite, usually fragments of shells

Mass wasting. A variety of processes by which large masses of earth material are moved by gravity, either slowly or quickly, from one place to another

Matrix. The small particles of a sediment or a sedimentary rock occupying the spaces between the larger particles that form the framework

Meander. One of a series of somewhat regular and looplike bends in the course of a stream, developed when the stream is flowing at grade, through lateral shifting of its course toward the convex sides of the original curves

Megascopic. Visible with the unaided eye or with a hand lens

Mesozoic. One of the grand divisions or eras of geologic time, following the Paleozoic and succeeded by the Cenozoic Era, and comprising the Triassic, Jurassic, and Cretaceous periods

Metamorphic rock. Any rock that has been recrystallized at a red heat, deep within the earth's crust; usually contains mineral crystals large enough to see and has a streaked or grainy appearance

Metamorphism. The process whereby sedimentary or igneous rocks have been altered by heat and pressure accompanying deep burial in the earth's crust

Metaquartzite. A quartzite formed by metamorphic recrystallization

Metasandstone. An altered or somewhat weakly metamorphosed sandstone

Metasiltstone. An altered or somewhat weakly metamorphosed siltstone

Mineral. A homogeneous, naturally occurring, solid substance of inorganic composition, consistent physical properties, and specified chemical composition

Mine spoils. The unused earth and rock excavated during mining activity; usually left in large mounds or piles around mining sites

Mississippian. The fifth period of the Paleozoic Era, from 310 to 350 million years ago

Mollusk. An invertebrate belonging to the phylum Mollusca, characterized by a nonsegmented body that is bilaterally symmetrical and by a radially or biradially symmetrical mantle and shell

Morphology. The observation of the form of lands

Mudflow. A flowage of heterogeneous debris lubricated with a large amount of water, usually following a former stream course

Nivation. Frost action and mass wasting beneath a snowbank

Oceanic rocks. Rocks making up the oceanic crust, usually darker and denser than those of the continents

Ordovician. The second period of the Paleozoic Era, from 430 to 500 million years ago

Ore deposit. An accumulation of minerals that can be mined at a profit; the minerals are termed ore minerals and the aggregate is termed ore

Orogeny. The process of the formation of mountains

Outcrop. Any exposure of bedrock

Overthrust. A type of fault in which an extensive slab of rock is moved across a nearly horizontal surface

Oxbow lake. A crescent-shaped body of water occupying the abandoned meander of a stream channel

Paleozoic. The first era of Phanerozoic geologic time, from 225 to 600 million years ago

Percolate. To pass through fine interstices, to filter, as in water through porous stones

Permeability. The capacity of a rock or soil for transmitting a fluid

Phanerozoic. Comprises Paleozoic, Mesozoic, and Cenozoic; eon of evident life

Phyllite. A metamorphic rock similar to schist but with grains so fine that they cannot be seen with the unaided eye

Phylum. A major unit in the taxonomy of animals, ranking above "class" and below "kingdom"

Planar failure. A landslide type (also called a block glide) in which movement of material occurs along a flat, inclined plane, usually along and parallel to bedding layers of rock

Plate. A large block of the earth's crust, separated from other blocks by midocean ridges, trenches, and collusion zones

Plateau. Any comparatively flat area of great extent and elevation

Plate tectonics. The movement of large blocks of the earth's crust over the surface of the globe

Plunge pool. A deep, circular lake occupying a basin beneath a waterfall

Plutonic. Pertaining to rocks and processes that occur deep within the earth's crust

Porosity. The ratio of the aggregate volume of voids in a rock or soil to its total volume, usually stated as a percentage

Precambrian. Geologic time before the Paleozoic Era

Pre-split face. In highway cutslopes, a smooth rock face developed by drilling closely spaced holes and lightly blasting to form a crack in the rock that connects the drilled holes

Proctor density. The mass of a substance (usually soil) derived by finding the weight and moisture content of a sample and measuring the volume occupied by the sample prior to removal

Profile. A drawing showing a vertical section of the ground along a surveyed line

Progradation. A seaward advance of the shoreline resulting from the near-shore deposition of sediments brought to the sea by rivers

Proto star. The energy and mass accretion disc at the beginning of star formation

Pyrite. A metallic, brass-colored iron-ore mineral (FeS_2), often called fool's gold, used as a source of sulfur for sulfuric acid

Pyrrhotite. A massive, magnetic, reddish-brown to bronze mineral of iron and sulfur

Quartz. A hard, glassy mineral, silicon dioxide (SiO_2), that is one of the most common rock-forming minerals, and a silicate group mineral

Quartzite. A sedimentary or metamorphic rock composed largely of quartz grains cemented by silica

Quaternary. The younger of the two geologic periods in the Cenozoic Era; includes all geologic time and deposits from the end of the Tertiary Period until and including the present; subdivided into the Pleistocene and Recent epochs or series

Radioactive. Containing atoms whose nuclei spontaneously disintegrate, releasing atomic particles and energy

Radioactive decay. The change of one element to another by the emission of charged particles from the nuclei of its atoms

Reclamation. The act or process of restoring a land or water environment to a state of usefulness

Recrystallization. Alteration of rocks whereby preexisting mineral grains are destroyed and new ones are formed, generally by increased heat and pressure; one of the metamorphic processes

Reef. A range or ridge of rocks lying at or near the surface of water; either moundlike or layered, built by sedimentary organisms such as corals, and usually enclosed in rock of a differing lithology

Regolith. The layer or mantle of loose, incoherent rock material, of whatever origin, that nearly everywhere forms the surface of the land and rests on the hard bedrocks

Relief. The actual physical shape, configuration, or general unevenness of the earth's surface; the difference in elevation between the high and low points of a land surface

Residuum. Soil formed in place by the disintegration and decomposition of rocks and the consequent weathering of the mineral materials

Rip-rap ditch. A drainage ditch lined with stones (rip-rap) in order to prevent soil erosion and to slow the flow of water

Roadway template. The configuration of a roadway bed, showing width of road surface, thickness of base stone gravel, kinds and thickness of asphalt or concrete pavement, width of shoulders, and locations of side drainage ditches

Rock arch. A naturally occurring landscape feature in which, through erosion and weathering, bedrock has been sculptured into a bridge-like arch

Rock bolt. A steel rod inserted in a drilled hole in a rock mass to anchor a plate or bar along the face of the rock

Rock pad. In highway construction, the use of blasted, broken rock in a 3-to-5-foot-thick layer, forming a "pad" over wet or soft areas; typically used for drainage

Rotational slide. A landslide type, usually in soil, in which the failure plane develops a rotational movement about an axis perpendicular to the failure plane

Sandstone. Sedimentary rock composed of sand grains

Saprolite. A soft, earthy, clay-rich, thoroughly decomposed rock formed in place by chemical weathering of igneous or metamorphic rocks

Saturated. Pertaining to rock or soil that has all its interstices filled with water

Scaling. The process of removing loose rock debris from an existing rock slope

Scarp. Cliff or steep break in a slope

Schist. A visibly crystalline metamorphic rock containing abundant mica or other cleavable minerals so aligned that the rock breaks regularly along the mineral grains

Sedimentary. Formed of fragments of other rocks transported from their source and deposited in water; also, transported in solution and deposited by chemical or organic agents

Seismic. Pertaining to, characteristic of, or produced by earthquakes or earth vibration

Sequence. A succession of stratified rocks

Serrated. Pertaining to the profile of small, step-like benches artificially cut into a hillside to reduce erosion, or to a topographic feature with a saw-toothed profile

Shale. Platy sedimentary rock formed from mud or clay, breaking easily parallel to the bedding

Shear strength. The internal resistance, in soil or rock, offered to shear (failure) stress

Silica. Silicon dioxide (SiO_2), occurring as quartz and as a major part of many other minerals

Siliceous. Containing abundant silica, especially as free silica rather than as silicates

Siltation. The deposition by a stream or river of graded particulates (sand, silt, clay) that tend to cover the stream bottom

Siltation fence. A fence composed of hay or straw bales, or synthetic fabric stretched along the boundary of an excavation, to reduce the flow of silt and mud from the excavated area

Siltstone. A sedimentary rock composed of mostly silt-sized particles (1/256 to 1/16 mm diameter)

Sinkhole. A large depression caused by collapse of the ground into an underlying limestone cavern

Slickensides. Grooves or scratches in rocks made by movement along a fault surface

Slide scar. A term used to describe the resulting surface following a landslide

Slip. The amount of movement on a fault measured on the fault surface: strike slip is the component of slip measured along the strike of the fault; dip slip is the component measured in the direction of the dip of the fault

Solifluction. The slow flowage from higher to lower ground of water-saturated masses of waste

Strata. Beds or layers of sedimentary rock

Stratified. Deposited in nearly horizontal layers of strata on the earth's surface

Stratigraphic. Pertaining to the composition, sequence, and correlation of stratified rocks

Strike. The direction or bearing of a horizontal line on a sloping bed, fault, or other rock surface

Subsidence. A sinking movement of soil and/or rock, in which there is no free side and surface material is displaced vertically downward with little or no horizontal component

Syncline. A fold in stratified rocks in which the strata on opposite sides usually dip inward toward each other

Synthetic geomembrane. An artificial, petroleum-based product, impervious to air and water, that is used in soil or rock applications

Talus. Blocks of rock pried loose by frost wedging; a type of colluvium, usually found at the base of a cliff or other steep slope

Talus slope. A landform underlain by talus

TDOT. Tennessee Dept. of Transportation

Tectonic. Pertaining to the larger structural features of the earth's crust and the forces that have produced them

Terraced deposit. A deposit of alluvial origin, composed of pebbles and cobbles in a sand or silt matrix, which is found on a relatively flat, elongate surface along the side of a valley; the deposit is formerly the alluvial floor of the valley

Tertiary. The earlier of the two geologic periods comprising the Cenozoic Era

Thrust fault. A fault, commonly of low dip, on which rocks have slid or have been pushed laterally over other rocks

Tidal flat. An extensive, nearly horizontal, marsh or barren tract of land that is alternately covered and uncovered by the rise and fall of the tide

Topography. The physical features of a district or region as represented on maps; the relief, contour, or shape of the land surface

Transgression. The gradual expansion of a shallow sea, resulting in the progressive submergence of land, as when sea level rises or land subsides

Translational slide. A landslide type in which movement occurs in a straight path

Trench. A long but narrow depression of the deep sea floor with relatively steep sides

Trimming. In highway construction and repair, a process of minor drilling and blasting of rock slopes to remove overhanging boulders and unstable rock

Unconformity. A buried surface of erosion or nondeposition separating younger rocks from older ones

Uvala. Large sinkhole karst formed by the coalescence of several doline sinks

Vein. A rock containing ore materials; usually a tabular mass

Water gap. A pass in a ridge that is still occupied by the stream that formed the gap

Water table. The upper surface of groundwater, below which soil and rock are saturated

Weathering. The processes that cause solid rock to decay into soil

Wedge failure. A landslide type in which movement occurs along the line of intersection of two inclined planes of discontinuity, resulting in a wedge-shaped scar on the landscape

Window. A hole produced by erosion through a thrust fault, exposing the underlying rocks

References

Aycock, J. H. 1981. Construction problems involving shale in a geologically complex environment—S.R. 32, Grainger Co., Tennessee. In *Proc. 32nd Annu. High. Geol. Symp.*, 36–58.

Barr, T. A. 1961. *Caves of Tennessee.* Tenn. Div. Geol. Bull. 64. 567 pp.

Beatty, S. M. 1978. *Why Not Walk?* Eastern National Park and Monument Association. 40 pp.

Blythe, E. W., W. T. McCutchen, and R. G. Stearns. 1975. Geology of Reelfoot Lake and vicinity. In *Field Trips in West Tennessee*, ed. R. G. Stearns, 64–76. Tenn. Div. Geol. Rep. Invest. 36.

Bowders, J., and S. Lee. 1990. Guide for selecting an appropriate method to analyze the stability of slopes on reclaimed surface mines. In *Proc. 1990 Mining and Reclamation Conf. and Exhib.*, vol. 1, 103–10. Morgantown: West Virginia Univ. Pub. Serv.

Bradham, W., and F. Caruccio. 1990. A comparative study of tailings analysis using acid/base accounting, cells, columns, and soxhlets. In *Proc. 1990 Mining and Reclamation Conf. and Exhib.*, vol. 1, 19–26. Morgantown: West Virginia Univ. Pub. Serv.

Brady, K., and R. J. Hornberger. 1989. Mine drainage prediction and overburden analysis in Pennsylvania. 13 pp. In *Proc. Annu. West Virginia Surface Mine Drainage Task Force Symp.*

Burgess, J. Z. 1977. *The Burgess history: The Tennessee pioneer, 1776–1976.* Cookeville, Tenn.: Crown Graphics. 300 pp.

Byerly, D. W. 1981. Evaluation of the acid drainage potential of certain Pre-Cambrian rocks in the Blue Ridge province. In *Proc. 32nd Annu. High. Geol. Symp.*, 174–85.

———. 1990. *Guidelines for handling excavated acid-producing materials.* Fed. High. Administration Spec. Doc. DOT FHWA-DF-89-0011. 81 pp.

Campbell, C. C. 1960. *Birth of a national park in the Great Smoky Mountains.* Knoxville: Univ. of Tennessee Press. 154 pp.

Caruccio, F. T., and G. Geidel. 1984. Induced alkaline recharge zones to mitigate acidic seeps. *Proc. Nat. Symp. on Surface Mining, Hydrology, Sedimentology and Reclamation,* 43–47.

Cherry, W. 1959. Geography of Tennessee. Dept. of Geography, Univ. of Tennessee, Knoxville. Mimeo.

Clark, M. G., et al. 1989. Central and Southern Appalachian geomorphology—Tennessee, Virginia, and West Virginia. In *Field Trip Guide Book* T150, 18–31. Washington, D.C.: American Geophysical Union.

Clark, M. G., P. T. Ryan, and E. C. Drumm. 1987. Debris slides and debris flows on Anakeesta Ridge, Great Smoky Mountains National Park, Tennessee. In *Landslides of Eastern North America.* U.S. Geol. Surv. Circ. 1008: 18–19.

Clifton, J. 1980. *Reelfoot and the New Madrid Quake.* Asheville, N.C.: Victor. 84 pp.

Coleman, Brenda D., and Jo Anna Smith. *Hiking the Big South Fork.* 2nd ed. Knoxville: Univ. of Tennessee Press, 1993.

Corgan, J. X. 1976. *Vertebrate fossils of Tennessee.* Tenn. Div. Geol. Bull. 77. 100 pp.

Corgan, J. X., and J. T. Parks. 1979. *Natural bridges of Tennessee.* Tenn. Div. Geol. Bull. 80. 102 pp.

Crawford, N. C. 1987. *The karst hydrogeology of the Cumberland Plateau escarpment of Tennessee.* Tenn. Div. Geol. Rep. Invest. 44, part 1. 43 pp.

Davidson, J. L. 1983. Paleoecological analysis of Holocene vegetation, Lake in the Woods, Cades Cove, Great Smoky Mountains National Park. M.S. thesis, Univ. of Tennessee, Knoxville. 100 pp.

Delcourt, P. A., and H. R. Delcourt. 1979. Late Pleistocene and Holocene distributional history of the deciduous forest in the southeastern United States. *Veröffentlichungen des Geobotanischen Institutes der ETH* 5: 79–107.

———. 1981. Vegetation maps for eastern North America: 40,000 yr. BP to the present. In *Geobotany II,* ed. R. Romans, 123–66. New York: Plenum.

———. 1985. Dynamic Quaternary landscapes of East Tennessee: an integration of paleoecology, geomorphology, and archaeology, Field Trip 7. In *Field Trips in the Southern Appalachians,* SE-GSA 1985, Univ. of Tennessee, Dept. of Geological Sciences, Studies in Geology 9: 191–220.

Denton, G. M. 1986. *Summary of the sedimentation studies of Reelfoot Lake, 1982–1986.* Nashville: Tenn. Dept. Health and Environ., Div. Water Manag. 34 pp.

Dunbar, C. O. 1919. *Stratigraphy and correlation of the Devonian of Western Tennessee.* Tenn. Div. Geol. Bull. 21. 127 pp.

Dunn, D. 1988. *Cades Cove: The life and death of a southern Appalachian community.* Knoxville: Univ. of Tennessee Press. 319 pp.

Ferm, J. C., R. C. Milici, and J. E. Eason. 1972. *Carboniferous depositional environments in the Cumberland Plateau of southern Tennessee and northern Alabama*. Tenn. Div. Geol. Rep. Invest. 33. 32 pp.

Floyd, R. J. 1965. *Tennessee Rock and Mineral Resources*. Tenn. Div. Geol. Bull. 66. 119 pp.

Fuller, M. L. 1912. *The New Madrid Earthquake*. U.S. Geol. Surv. Bull. 494. 119 pp.

Glass, B. P. 1982. *Introduction to planetary geology*. New York: Cambridge Univ. Press. 469 pp.

Hatcher, R. D., and P. J. Lemiszki. 1991. Western Great Smoky Mountains windows: the foothills duplex. In *Structure of Southern Appalachian Foreland Fold-Thrust Belt in Tennessee* (Field Trip Guide for Tenn. Div. Geol. Field Trip, Nov. 12–13, 1991), 1–122.

Hutson, S. S. 1989. USGS and division of superfund to investigate potential contamination at the well field supplying the city of Alamo in Crockett County, West Tennessee. *Water Resources in Tennessee: Current Conditions* (U.S. Geol. Surv., Water Resources. Div., Nashville) 3, no. 3: 7.

Kemmerly, P. R. 1980. *Sinkhole collapse in Montgomery County, Tennessee*. Tenn. Div. Geol. Environ. Geol. Series , no. 6. 42 pp.

King, P. B. 1964. *Geology of the central Great Smoky Mountains, Tennessee*. U.S. Geol. Surv. Prof. Pap. 349-C. 148 pp.

King, P. B., R. B. Neuman, and J. B. Hadley. 1968. *Geology of the Great Smoky Mountains National Park, Tennessee and North Carolina*. U.S. Geol. Surv. Prof. Pap. 587. 23 pp.

Kirkeminde, P. B. 1977. *Cumberland homesteads*. Crossville, Tenn.: Bookhart. 79 pp.

Klemens. T. L. 1992. *Infrastructure*. Des Plaines, Ill.: Cahners.

Klepser, H. J. 1967. *Historical geology laboratory manual*. Dept. of Geology, Univ. of Tennessee, Knoxville. 85 pp.

Knoll, A. H. 1991. End of the Proterozoic Eon. In *Sci. Am*. 265, no. 4 (Oct.): 64–73.

Kopal, Z. 1973. *The solar system*. New York: Oxford Univ. Press. 152 pp.

Lane, C. F. 1952. Grassy Cove, a uvala in the Cumberland Plateau, Tennessee. *J. Tenn. Acad. Sci*. 27: 291–95.

Laurence, R. A., and A. R. Palmer. 1963. *Age of the Murray Shale and Hesse Quartzite on Chilhowee Mountain, Blount County, Tennessee*. U.S. Geol. Surv. Prof. Pap. 475-C, art. 73, pp. C53–54.

Leech. W. M. 1956. *Report of the State Highway Commissioner of Tennessee for the Biennium ending June 30, 1956*. Nashville: Tenn. Dept. High. 202 pp.

———. 1958. *Report of the State Highway Commissioner of Tennessee for the Biennium ending June 30, 1958*. Nashville: Tenn. Dept. High. 193 pp.

Luther, E. T. 1977. *Our restless earth*. Knoxville: Univ. of Tennessee Press. 94 pp.

Mainfort, R. C. 1980. *Archaeological investigations at Fort Pillow State Historic Area: 1976–1978*. Tenn. Dept. Conserv., Div. Archaeology, Res. Ser., no. 4. 198 pp.

Manning, R., and S. Jamieson. 1990. *The South Cumberland and Fall Creek Falls*. Norris, Tenn.: Laurel Place. 103 pp.

Marcher, M. V., and R. G. Stevens. 1962. *Tuscaloosa Formation in Tennessee.* Tenn. Div. Geol. Rep. Invest. 17. 22 pp.

Matthews, L. E. 1971. *Descriptions of Tennessee caves.* Tenn. Div. Geol. Bull. 69. 150 pp.

McIntire, S. C., J. W. Naney, and J. C. Lance. 1986. *Sedimentation in Reelfoot Lake.* Nashville: Tenn. Dept. Health and Environ., Div. Water Manag. 58 pp.

McLaughlin, R. E. 1973. Observations on the biostratigraphy and stratigraphy of Knox County, Tennessee and vicinity. In *The Geology of Knox County,* Tenn. Div. Geol. Bull 70, 25–62.

McLaughlin, R. E., and H. L. Moore. 1975. Paleoecologic analysis of the Coon Creek fauna. In *Abstracts with Programs* (24th Annu. Southeastern Section Meeting, Geol. Soc. Am.) 7, no. 4: 517.

McMaster, W. M. 1991. Groundwater systems of the Oak Ridge area. In *Structure of the Southern Appalachian Foreland Fold-Thrust Belt in Tennessee: Comparison of Styles, Mesofabrics, Geometry, and Detachment Levels in a Thrust Dominated Orogon* (Field Trip Guide for Tenn. Div. Geol. Field Trip, Nov. 12–13, 1991), ed. R. D. Hatcher and P. J. Lemiszki, 85–88.

Milici, R. C. 1967. The physiography of Sequatchie Valley and adjacent portions of the Cumberland Plateau, Tennessee. *Southeast. Geol.* 8, no. 4 (Dec.): 179–93; rpt. 1968 as Tenn. Div. Geol. Rep. Invest. 22.

Miller, R. A. 1974. *The geologic history of Tennessee.* Tenn. Div. Geol Bull. 74. 63 pp.

Miller, R. A., and S. W. Maher. 1972. *Geologic evaluation of sanitary landfill sites.* Tennessee. Tenn. Div. Geol. Environ. Geol. Ser., no. 1. 38 pp.

Mitchell, B. J. 1991. Mid-continent earthquake zones—lessons from New Madrid, Missouri. *Earthquakes and Volcanoes* 22, no. 3: 120–23.

Moore, G. K., C. R. Burchett, and R. H. Bingham. 1969. *Limestone hydrology in the upper Stones River basin, Central Tennessee.* Tenn. Div. Water Resources. 58 pp.

Moore, H. L. 1974. A systematic and paleoecologic review of the Coon Creek fauna. M.S. thesis, Univ. of Tennessee, Knoxville. 187 pp.

———. 1976. The piping phenomenon. In *Proc. 27th Annu. High. Geol. Symp.,* 112–30.

———. 1980. Karst problems along Tennessee highways, an overview. In *Proc. 31st Annu. High. Geol. Symp.,* 1–28.

———. 1984. Geotechnical considerations in the location, design, and construction of highways in karst terrain—The Pellissippi Parkway Extensions, Knox-Blount counties, Tennessee. In *Sinkholes: Their Geology, Engineering, and Environmental Impact,* Proc. 1st Multidisciplinary Conf. on Sinkholes, 385–89.

———. 1986. Wedge failures along Tennessee highways in the Appalachian region: Their occurrence and correction. *Bull. Assoc. Eng. Geol.* 23, no. 4 (Nov.): 441–60.

———. 1987a. Karst vs. highway ditchlines in East Tennessee. In *Proc. 38th Annu. High. Geol. Symp.,* 117–23.

———. 1987b. Sinkhole development along "untreated" highway ditchlines in East Tennessee. In *Proc. 2nd Multidisciplinary Conf. on Sinkholes and the Environmental Applications*, ed. B. Beck and W. Wilson, 115–19. New York: A. A. Balkema.

———. 1988a. Oriented Pre-split for controlling rock slides. In *Proc. 37th Annu. High. Geol. Symp.*, 236–50.

———. 1988b. *A roadside guide to the geology of the Great Smoky Mountains National Park*. Knoxville: Univ. of Tennessee Press. 178 pp.

———. 1988c. Treatment of karst along Tennessee highways. In *Geotechnical Aspects of Karst Terrains*, Am. Soc. Civil Eng. Geotechnical Spec. Pub. 14., 133–48.

———. 1990. Rock fall mitigation along I-40, Cocke and Cumberland counties, Tennessee. In *Proc. 41st Annu. High. Geol. Symp.*, 1–10.

———. 1992. The use of geomembranes for mitigation of pyritic rock. In *Proc. 43rd Annu. High. Geol. Symp.*, 441–63.

Moore, H. L., and D. Amari. 1987. Sinkholes and gabions: A solution to the solution problem. In *Proc. 2nd Multidisciplinary Conf. on Sinkholes and the Environmental Implications*, ed. B. Beck and W. Wilson, 305–10. New York: A. A. Balkema.

Moulton, D. W. 1960. *Report of the State Highway Commissioner of Tennessee for the Biennium ending June 30, 1960*. Nashville: Tenn. Dept. High. 191 pp.

———. 1962. *Report of the State Highway Commissioner of Tennessee for the Biennium ending June 30, 1962*. Nashville: Tenn. Dept. High. 213 pp.

Murlless, D., and C. Stallings. 1973. *Hiker's guide to the Smokies*. New York: Sierra Club. 375 pp.

Neuman, R. B., and W. H. Nelson. 1965. *Geology of the western part of the Great Smoky Mountains, Tennessee*. U.S. Geol. Surv. Prof. Pap. 349-D. 81 pp.

Newton, J. G. 1976. *Early detection and correction of sinkhole problems in Alabama, with a preliminary evaluation of remote sensing applications*. Montgomery, Ala. High. Dept. HPR Rep. 76, Res. Proj. 930-070. 83 pp.

Obermeier, S. F. 1989. *The New Madrid earthquakes: An engineering-geologic interpretation of relict liquefaction features*. U.S. Geol. Surv. Prof. Pap. 1336-B. 114 pp.

Pack, D. M. 1964. *Report of the State Highway Commissioner of Tennessee for the Biennium ending June 30, 1964*. Nashville: Tenn. Dept. High. 137 pp.

———. 1966. *Report of the State Highway Commissioner of Tennessee for the Biennium ending June 30, 1966*. Nashville: Tenn. Dept. High. 173 pp.

Parks, W. S. 1990. Shelby County landfill groundwater sampling completed. *Water Resources in Tennessee: Current Conditions* (U.S. Geol. Surv., Water Resources. Div., Nashville) 4, no. 3: 9.

Parks, W. S., and R. W. Lounsbury. 1975. Environmental geology of Memphis, Tennessee. In *Field trips in West Tennessee*, ed. R. G. Stearns, 35–63. Tenn. Div. Geol. Rep. Invest. 36.

Peacock, B. G. 1973. Reelfoot Lake State Park. *Tenn. Hist. Q.* 32, no. 3 (Fall): 205–32.

Price, B. 1991. Tennessee's mystery craters. *Tenn. Conserv.* 57, no. 5 (Sept.–Oct.): 22–26.

Quinlan, J. F. 1983. Groundwater hydrology and geomorphology of the Mammoth Cave region, Kentucky, and of the Mitchell Plain, Indiana. In *Field Trips in Midwestern Geology* (Geol. Soc. Am. and Ind. Geol. Survey), vol. 2, ed. R. H. Shaver and J. A. Sunderman, 1–85.

Rae, J. B. 1971. *The road and the car in American life.* Cambridge, Mass.: MIT Press. 390 pp.

Reeves, R., and P. Somers. 1991. Cedar glades of Middle Tennessee. *Tenn. Conserv.* 57, no. 3 (May–June): 11–14.

Rodgers, J. 1953. *Geologic map of East Tennessee with explanatory text.* Tenn. Div. Geol. Bull. 58, part 2. 162 pp.

Royster, D. L. 1973. Highway landslide problems along the Cumberland Plateau in Tennessee. *Bull. Assoc. Eng. Geol.* 10, no. 4 (Fall): 255–87.

———. 1977. Some observations on the use of horizontal drains in the correction and prevention of landslides. In *Proc. 28th Annu. High. Geol. Symp.*, 137–45,

———. 1979. Landslide remedial measures. *Bull. Assoc. Eng. Geol.* 16, no. 2 (May): 301–52.

———. 1984. The use of sinkholes for drainage. In *Trans. Res. Rec.* (Washington, D.C.), no. 978, 18–25.

Russell, E. E. 1965. Stratigraphy of the outcropping Cretaceous beds below the McNairy Sand in Tennessee. Ph.D. diss., Univ. of Tennessee. 197 pp.

Russell, E. E., and W. S. Parks. 1975. *Stratigraphy of the outcropping Cretaceous, Paleocene, and lower Eocene of Western Tennessee.* Tenn. Div. Geol. Bull. 75. 125 pp.

Safford, J. M. 1869. *Geology of Tennessee.* Nashville: Bur. Agri. and Commer. 550 pp.

Schoner, A. E. 1985. The Clinch Sandstone (Lower Silurian) along Highway 25E at Beans Gap. In *The geologic history of the Thorn Hill Paleozoic section (Cambrian-Mississippian), Eastern Tennessee*, ed. K. R. Walker, Univ. of Tennessee Dept. of Geological Sciences, Studies in Geology 10, 100–110.

Sellers, D. 1991. Dunbar Cave: Social playground to natural habitat. *Tenn. Conserv.* 57, no. 2 (Mar.–Apr.): 8–11.

Sobek, A. A., et al. 1978. *Field and laboratory methods applicable to overburdens and minesoils.* EPA Pub. EPA-600/2-78-054. 203 pp.

Spencer, E. W. 1962. *Basic concepts of historic geology.* New York: Crowell. 504 pp.

Stearns, R. G. 1957. *Cretaceous, Paleocene, and Lower Eocene geologic history of the northern Mississippi Embayment.* Geol. Soc. Am. Bull. 68: 1077–100.

———. 1975. *Field trips in West Tennessee.* Tenn. Div. Geol. Rep. Invest. 36. 82 pp.

Stearns, R. G., and C. A. Armstrong. 1955. *Post-Paleozoic stratigraphy of Western Tennessee and adjacent portions of the upper Mississippi Embayment.* Tenn. Div. Geol. Rep. Invest. 2. 29 pp.

Stearns, R. G., E. W. Blythe, and M. L. Hoyal. 1989. *Guidebook of geology and related history of the Chickasaw Bluffs and Mississippi River Valley in West Tennessee.* Tenn. Div. Geol. Rep. Invest. 45. 44 pp.

Stearns, R. G., and M. V. Marcher. 1962. *Late Cretaceous and subsequent structural development of the northern Mississippi Embayment area.* Geol. Soc. Am. Bull. 73: 1387–94.

Stearns, R. G., and J. M. Wilson. 1971. *Hydrology, geology and erosion by leaching in Skillman Basin on the Western Highland Rim, Lawrence Co.* Tennessee. Tenn. Div. Water Resources Res. Ser., no. 3. 33 pp.

Stiller, A. H. 1982. A method for prevention of acid mine drainage. 11 pp. In *Proc. 3rd West Virginia Surface Mine Drainage Task Force Symp.*

Sullivan, L. A. Little Fort Pillow. 1990. *Tenn. Conserv.* 56, no. 2 (Mar.–Apr.): 25–26.

Templeton, T. R., and B. C. Spencer. 1980. *Earthquake data for Tennessee and surrounding areas.* Tenn. Div. Geol. Environ. Geol. Ser., no. 8. 63 pp.

Tennessee Dept. of Transportation (TDOT). 1976. *Interstate Route 40 Environmental Impact and Section 4F Statement, Memphis, Tenn.* [FHWA-TN-EIS-76-03-D] Nashville: Tenn. Dept. Transp., Bur. Planning and Programming. 443 pp.

———. 1977. *Briefing paper of the proposal for completion of I-40, Memphis, Tenn. from Claybrook St. to Bon Air St.* Nashville: Tenn. Dept. Transp., Bur. Planning and Programming. 66 pp.

Tennessee State Highway Dept. (TSHD). 1959. *History of the Tennessee Highway Department.* Nashville: State of Tennessee. 126 pp.

Torbert, J., and J. Burger. 1990. Guidelines for establishing productive forest land on reclaimed surface mines in the Central Appalachians. In *Proc. 1990 Mining and Reclamation Conf. and Exhib.*, vol. 1, 273–78. Morgantown: West Virginia Univ. Pub. Serv.

Tozer, D. C. 1977. The thermal state and evolution of the earth and terrestrial planets. *Sci. Prog.*, no. 64: 1–28.

Trolinger, W. O. 1975. Construction of the I-40 reinforced earth embankment. In *Proc. 56th Annu. Tennessee High. Conf.*, Bull. No. 41 (Jan. 1), 24–32.

Tyler, P. 1957. *American highways today.* New York: H. W. Wilson. 204 pp.

Unrug, R., and S. Unrug. 1990. Paleontological evidence of Paleozoic age for the Walden Creek Group, Ocoee Supergroup, *Tenn. Geol.* 18: 1041–45.

Unrug, R., S. Unrug, and S. L. Palmes. 1991. Carbonate rocks of the Walden Creek Group in the Little Tennessee River Valley: Modes of occurrence, age and significance for the basin evolution of the Ocoee Supergroup. In *Studies of Precambrian and Paleozoic stratigraphy in the Western Blue Ridge* [Carolina Geol. Soc.], 27–38.

U.S. Dept. of Transportation. 1976. *America's highways, 1776–1976: A history of the Federal-Aid Program.* Washington, D.C.: U.S. Dept. of Transportation, Federal Highway Administration. 553 pp.

Vandover, J. D. 1989. Stratigraphy of pilot tunnel through Cumberland Mountain (Kentucky-Tennessee). M.S. thesis, Eastern Kentucky Univ. 78 pp.

Vaughn, L. M. 1927. *Just the little story of Cumberland Gap.* Middlesborough, Ky.: Cumberland Gap National Historic Park Association. 16 pp.

Wade, B. 1926. *The fauna of the Ripley Formation on Coon Creek, Tennessee.* U.S. Geol. Surv. Prof. Pap. 137. 272 pp.

Walker, J. D. 1990. The sedimentology and stratigraphy of the Chilhowee Group (Uppermost Proterozoic to Lower Cambrian) of Eastern Tennessee and Western North Carolina: The evolution of the Laurentian-Iapetus margin. Ph.D. diss., Univ. of Tennessee, Knoxville. 274 pp.

Walker, K. R., ed. 1985. *The geologic history of the Thorn Hill Paleozoic section (Cambrian-Mississippian), Eastern Tennessee.* Univ. of Tennessee Dept. of Geological Sciences, Studies in Geology 10 (SE-GSA 1985 Field Trip 6). 128 pp.

Walker, K. R., and K. F. Ferrigno. 1973. Holston Formation—An insight into the three-dimensional relationships between reef-core and reef flank facies. In *The Geology of Knox County,* Tenn. Div. Geol. Bull 70, 131–38.

Walker, P. E. 1929. *Illustrated history of Reelfoot Lake.* Ridgely, Tenn.: Paul Walker. 25 pp.

Wilson, C. W. 1980. *Geology of Cedars of Lebanon State Park and Forest and vicinity in Wilson County, Tennessee.* Tenn. Div. Geol. State Park Ser., no. 1. 19 pp.

Wilson, C. W., and R. G. Stearns. 1968. *Geology of the Wells Creek Structure.* Tennessee. Tenn. Div. Geol. Bull. 68. 236 pp.

Wilson, J. T. 1963. Continental drift. *Sci. Am.* (Apr.): 41–55. Rpt. in *Continents adrift: Readings from* Scientific American. San Francisco: W. H. Freeman, 1972.

Wilson, R. L. 1981. *Guide to the geology along the interstate highways in Tennessee.* Tenn. Div. Geol. Rep. Invest. 39. 79 pp.

Wood, H. O., and F. Neumann. 1931. Modified Mercalli Intensity Scale of 1931. *Seismol. Soc. Am. Bull.* 21: 277–83.

Zak, J., and S. Visser. 1990. Soil microbial processes and dynamics: Their importance to effective reclamation. In *Proc. 1990 Mining and Reclamation Conf. and Exhib.,* vol. 1, 83–84. Morgantown: West Virginia Univ. Pub. Serv.

Ziemkiewicz, P. 1990. Advances in the prediction and control of acid mine drainage. In *Proc. 1990 Mining and Reclamation Conf. and Exhib.,* vol. 1, 51–54. Morgantown: West Virginia Univ. Pub. Serv.

Index

References to illustrations are printed in **boldface.**

A Geologic Trip across Tennessee by Interstate 40 was designed and composed by Sheila Hart for the University of Tennessee Press on the Apple Macintosh using Microsoft Word® and Aldus PageMaker®. Maps were created by the University of Tennessee Cartographic Laboratory under the supervision of Will Fontanez, using Aldus Freehand®. Linotronic camera pages were generated by The TypeCase. The book is set in Palatino with Benguiat Bold for display and is printed on recycled 60-lb. Thor White. Manufactured in the United States of America by Braun-Brumfield, Inc.